Learning Materials in Biosciences

Learning Materials in Biosciences textbooks compactly and concisely discuss a specific biological, biomedical, biochemical, bioengineering or cell biologic topic. The textbooks in this series are based on lectures for upper-level undergraduates, master's and graduate students, presented and written by authoritative figures in the field at leading universities around the globe.

The titles are organized to guide the reader to a deeper understanding of the concepts covered.

Each textbook provides readers with fundamental insights into the subject and prepares them to independently pursue further thinking and research on the topic. Colored figures, step-by-step protocols and take-home messages offer an accessible approach to learning and understanding.

In addition to being designed to benefit students, Learning Materials textbooks represent a valuable tool for lecturers and teachers, helping them to prepare their own respective coursework.

More information about this series at http://www.springer.com/series/15430

Cerian Ruth Webb
Mirela Domijan

Introduction to MATLAB® for Biologists

Cerian Ruth Webb
Department of Plant Sciences
University of Cambridge
Cambridge, UK

Mirela Domijan
Department of Mathematical Sciences
University of Liverpool
Liverpool, UK

ISSN 2509-6125 ISSN 2509-6133 (electronic)
Learning Materials in Biosciences
ISBN 978-3-030-21336-7 ISBN 978-3-030-21337-4 (eBook)
https://doi.org/10.1007/978-3-030-21337-4

This Springer imprint is published by the registered company Springer Nature Switzerland AG
The registered company address is: Gewerbestrasse 11, 6330 Cham, Switzerland

For Alys, my beautiful little mathematician
Cerian

For Emma, Bella and Strawberry
Mirela

Preface

Our aim in writing this book is to provide a self-guided introduction of MATLAB® for those who have little or no prior experience of coding and using structured programming languages. The book is part of a series of books aimed at bioscientists and so we have attempted to steer away from the usual engineering and physical examples used in many coding books and instead use examples which are hopefully more intuitive for those working in the life sciences.

This book is based on a course which we have developed and taught for several years at the University of Cambridge which is open to researchers ranging from PhD students to group leaders. The course attendees come with a wide variety of experiences from disciplines ranging from psychology and medical imaging, through to microbiology. Some come with experience of using specialist packages, others have failed to get to grips with alternative structured programming languages and many have no experience of programming at all. Our aim is to provide attendees, and now you, not just with the essential skills to use MATLAB, but also with some more advanced skills which will allow you to automate processes so that you can use MATLAB efficiently in your own research.

A core aim of the book is to give you the confidence to navigate around MATLAB so that you can access the documentation for the functions and toolboxes relevant to your specialist areas of research. We use MATLAB for a range of applications in our own work and happily admit that we make frequent use of the MATLAB documentation and online resources to work out how to solve programming issues and to explore unfamiliar functions. In writing this book, we had to select from the vast array of features available in MATLAB and finally have chosen to focus on functions and features that we use a lot and we hope that these will provide ideas on how you can tailor MATLAB to your own work.

We have avoided the use of any of the specialist MATLAB toolboxes for two reasons. First, we wanted all chapters of the book to be accessible to every reader regardless of what level of license you have; and second, we have found that the wide range of applications of MATLAB to biological systems mean that when we write extra worksheets on particular toolboxes, they are often only relevant to a small proportion of attendees and, by extension, readers. To include a chapter for each possible area of research would fill several more textbooks. We hope that working through this book will make it easier for you to work out how to use specific toolboxes by accessing the resources on the MathWorks® website.

Cerian Ruth Webb
Cambridge, UK

Mirela Domijan
Liverpool, UK
November 2018

Acknowledgements

This book is strongly influenced by the authors' experience developing and teaching a 2-day course in MATLAB® at the University of Cambridge. We would like to thank Nik Cunniffe and James Cox who contributed to the content of the original course and Paul Judge, David Judge and Gabriella Rustici who provide the facilities and support required to run the course. We would also like to thank the various post-docs who have helped on the course, including Ciara Dangerfield, Laura Hall and Andy Harrison. Thanks also to Clare Allen and Donovan Hide for their useful comments on this book.

Structure of the Book

We have arranged the book based on our experience of teaching a 2-day course on MATLAB. We do not, however, expect you to work through the whole book and absorb everything in 2 days. The first six chapters cover the basics of MATLAB. Once you have worked through these, you may wish to skip any of the other chapters. However, to get the most out of the package, we recommend going back to any chapters you have missed.

It is difficult to write the book in strictly linear order, and it has been necessary to sometimes include coding features before we have given a detailed introduction of them. Where we have done this, we have tried to give a pointer to the relevant chapter. This does not mean you need to skip to that chapter yet, but it is just to alert you that more detail is provided elsewhere in the book. It does not matter if you do not fully understand loops, logical statements and conditional statements the first time you encounter them. They are included in early examples so that we could both provide more interesting coding examples and help you develop an intuitive feeling for these programming constructs before their more formal introduction.

Contents

About the Authors

Dr. Cerian Ruth Webb
is an associate lecturer at Newnham College, University of Cambridge, and a researcher in the Epidemiology and Modelling group in the Department of Plant Sciences, University of Cambridge. She has lectured courses in Mathematical Biology, MATLAB® and infectious disease dynamics and has worked on mathematical models for a range of biological systems in veterinary and plant epidemiology.

Dr. Mirela Domijan
is a lecturer in Applied Mathematics in the Department of Mathematical Sciences at the University of Liverpool. She has lectured courses in Mathematical Biology and Systems Biology, and her research focuses on the application of mathematics to plant biology.

Getting Started

C. R. Webb, M. Domijan, *Introduction to MATLAB® for Biologists*,
Learning Materials in Biosciences, https://doi.org/10.1007/978-3-030-21337-4_1

"MATLAB has come a long way since the simple calculator that started it all. It is a living ecosystem supporting all aspects of technical computing. We will continue to strengthen existing features as we carefully add new ones. Our goals are always ease of use, power and speed". *Cleve Moler. MathWorks®*

1.1 Introduction

The name MATLAB® is an acronym for Matrix Laboratory and highlights its origins as a package of routines for matrix algebra. The first commercial version of MATLAB was released in 1984 with the MATLAB desktop version released in 2000. MATLAB is continuously evolving to reach a wider audience, and in 2016 a Live Editor interface was released enabling users to produce interactive documents which combine descriptive text, images and MATLAB code.

MATLAB has developed into an accessible structured programming language with over 60 toolboxes available which contain specialized sets of functions. At first glance the package may appear more targeted at engineers, mathematicians and physicists. However, MATLAB is a powerful and user-friendly package which will make it easier for you to produce publication standard plots, explore and analyse your data, conduct image analysis and solve mathematical models of biological systems. A key advantage of MATLAB is that you can create code which will allow you to reproduce your results and to perform the same set of commands on multiple sets of data with minimal effort.

1.2 Who Is This Book for?

This book is aimed at anyone who wants to learn how to use MATLAB regardless of prior experience with specialist packages and programming languages. Our target audience are natural scientists; however, we do not assume any subject-specific knowledge and hope the book will be useful to readers from a wide range of disciplines. The book is written for self-guided study but could also be used as a complement to a taught course or for group study.

The book aims to take you from your first introduction to MATLAB to a point where you can write your own functions and use coding constructs such as loops and conditional statements. Users with a high-level knowledge of related packages and programming may want to skim through the earlier chapters; however, it is still worth taking time to familiarize yourself with MATLAB-specific syntax. You may have started this book thinking that you just want to use MATLAB to make publication-quality figures or to conduct basic data analysis, but we hope that by working through the book, you will find other ways in which you could use MATLAB to assist in your research.

1.3 Installation of MATLAB®

MATLAB is a commercial software with a variety of purchasing options depending on whether you are working in industry, a student, working for an academic institute or want to purchase a licence for home use. Details of how to purchase MATLAB and how to download the product can be found on the MathWorks® website ► https://www.mathworks.com. If you do not currently have a licence and want to try out MATLAB, then you can download a 30-day trial licence from the same site. If you work in academia, it is worth checking whether your university has purchased a site-wide licence that you can download onto your own computer.

When you download MATLAB onto your computer, you will be given the option of installing any toolboxes included in your licence. In this book we have tried to avoid using toolboxes; however, many biological researchers find the Statistics and Machine Learning Toolbox™, Bioinformatics Toolbox™ and the Image Processing Toolbox™ useful, so if you have these as part of your licence, we would recommend downloading them ready to explore once you have worked through the book.

Old versions of MATLAB were referred to by release number; however, since 2006 the biannual releases are more readily identified by the year of release followed by a or b, for example, 2018a and 2018b. This book was written using MATLAB 2018b. Most of the commands used in this book should be compatible with releases which are a couple of years older and with newer versions as they become available. The main differences are the recent introduction of Live Scripts (2016a onwards) and Live Functions (2018a onwards) and refinements in the MATLAB interface which may mean that some shortcuts do not exist or can be found in a different place in older versions. It is likely that more features will be added to Live Scripts and Live Functions in future releases of MATLAB. If your licence includes updates (software maintenance service), we recommend updating to the latest version.

1.4 Starting MATLAB®

Once installed on your computer, MATLAB can be started in the same way as other programs on your computer, for example, if you are using Microsoft Windows type MATLAB into the search bar and if MATLAB is installed, the package will come up in the list of nearest matches. Depending on how fast your computer is, MATLAB may take a little while to load: you will first see a window that displays the MATLAB logo and details of the release number, and then the MATLAB desktop window will launch.

If this is the first time you have opened MATLAB, the interface should look something like ◘ Fig. 1.1.

The MATLAB interface consists of a toolstrip along the top of the interface and, on first start-up, three windows. As you work through the book, you will discover that you can add other windows to the interface, but for now you should have the following windows:

`Current Folder`: This window allows you to navigate around the files on your computer. Double-clicking on any folder in this window will change the working directory and hence the default location of anything you save where you have not specified a file path.

`Command Window`: The Command Window contains some introductory text followed by >>; this is the command line prompt. This is where you should enter the commands given in ► Chaps. 2 and 3. In ► Chap. 4 you will learn how to open and use the Editor window to run groups of commands efficiently.

`Workspace`: This shows all the variables which you have created or imported from files and which can be currently accessed by any commands you type into the Command Window.

Only one window is active at a time, and, for some window types, the toolstrip displayed will change according to which window is active. The title bar of the active window will be dark blue. To change the active window, just click in the title bar of the window you want to be active or anywhere in that window.

1

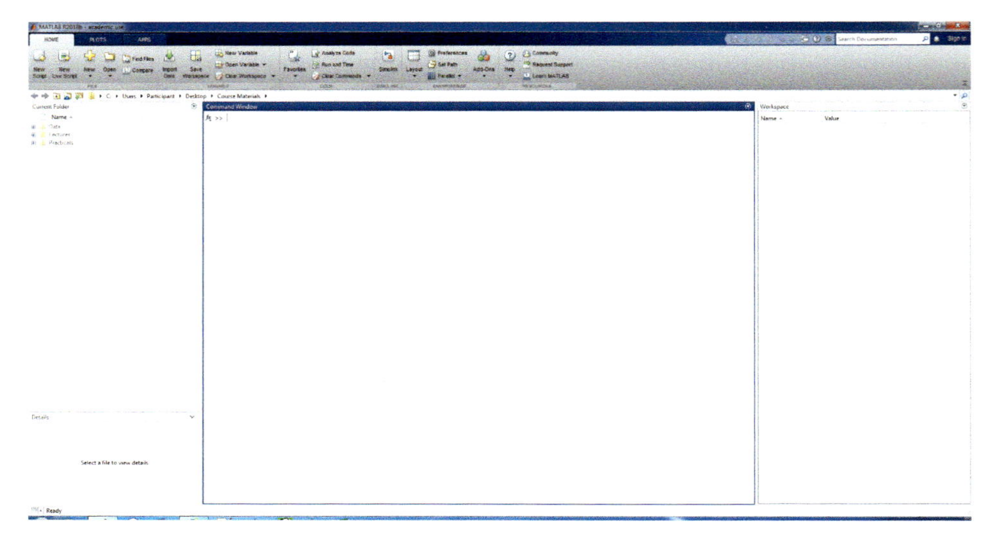

Fig. 1.1 Example of the MATLAB® Desktop

Fig. 1.2 Entering code into the Command Window

```
>> a = 2+3

a =

        5

fx >> |
```

If the Command Window is active, you should see a flashing cursor by the command prompt. Any commands should be typed into the Command Window (or the Editor which we introduce in ▶ Chap. 4). Pressing ENTER after you have finished the command will execute it. We do not include the command prompt in the code examples in this book as we encourage you to use the Editor once you have got used to the basics of MATLAB. Every time a new line of code is started, you should first press ENTER to execute the previous line so that you have a fresh command prompt. For example, if we present you with:

```
a = 2+3
b = 6.*10
```

We would expect you to start by typing, at the command prompt, `a = 2+3`, and then press ENTER. This will produce the output shown in Fig. 1.2 and a new command prompt at which you would type in `b = 6.*10`.

1.5 Finishing a MATLAB® Session

To end a MATLAB session and close the program, type

```
quit
```

When you next open MATLAB, you will see the same layout as you left the program in when you last closed it. Any variables not cleared from the workspace will still be there, and you will also be able to reuse previous commands if you have not cleared the Command History (see ▶ Chap. 2).

1.6 MATLAB® Mobile App and Online

Like the desktop application of MATLAB, when you purchase a licence for MATLAB, you will also be able to download the MATLAB® Mobile™ app onto your iPhone, iPad or Android mobile device. This app can be used as a stand-alone program or can be connected to a MATLAB session running on MathWorks® Cloud or on your computer. A useful feature of the app is that it can be used to collect data from device sensors such as the accelerometer and GPS.

If you have a licence for the current version of MATLAB, then you will be able to access MATLAB® Online™ – this gives you access to MATLAB from any standard web browser by logging into your MathWorks account without the need to have MATLAB installed on the computer. Be aware that there are some aspects of MATLAB not currently available in MATLAB Mobile and MATLAB Online (for full details see the MathWorks website).

1.7 How to Get the Most Out of the Book

The book includes many sections of code for you to try out – these are mostly presented in boxes in Courier font. We have not included the command prompt you will see on the screen, but you should assume that if we have started a new line then you should press ENTER and start a new line too. To get the most out of the book, you should ideally type the code in yourself and look at the output – if you are using the e-book, you can of course copy the code directly into MATLAB; however, be aware that this shortcut can reduce learning opportunities. Do not worry if you make a mistake copying the code into MATLAB – in fact a great way to learn how to use a package is to make errors and then work out what went wrong and how to fix it. It is worth taking time as you work through the book to explore what happens if you alter our code and to try out related calculations. You can answer questions such as "do I need to put a space here?" for yourself by seeing whether an error is generated if you don't.

We have not included lists of questions in the book but instead recommend that you make use of Cody™, ► https://www.mathworks.com/matlabcentral/cody, to test your understanding. This has a major advantage over providing questions and solutions in the book, in that it automatically checks your code against test cases and allows you to access other users' solutions, so you can see alternative approaches to solving the same problem. More detail of how to use Cody is given in ► Chap. 5.

Take-Home Messages

This book aims to teach you the basics of MATLAB® so that you can apply it to your own work. The most important piece of advice we can give you is not to skip ► Chap. 5 'Accessing Help': read it, use it and try out the problems in Cody™ – we can't include every command and option in this book. MathWorks® don't produce hard copies of the manuals any more – there's just too much: the last copy one of us had took up a whole shelf. Experiment and explore: input the commands we have provided into MATLAB, but then see what happens if you change them a little. Read the error and warning messages to see how you can improve your code.

The Basics Using the Command Line

C. R. Webb, M. Domijan, *Introduction to MATLAB® for Biologists*,
Learning Materials in Biosciences, https://doi.org/10.1007/978-3-030-21337-4_2

What You Will Learn in This Chapter

In this chapter, you will learn the basics of MATLAB® using the command line. The notation for basic mathematical operators is introduced and the role of different types of parentheses summarized. MATLAB records your command history, and we will show you how to access this so that you can repeat commands without having to retype them and how to add comments to your code. You will discover how floating-point numbers can lead to rounding errors in computational calculations. You will learn how to assign calculated values to variables, the rules for naming variables, how to use those variables in further calculations and how to erase them when they are no longer needed.

2.1 Finding the Command Line

The Command Line refers to the prompt in the Command Window. This window should appear in the default layout when you open MATLAB®. If you have accidentally closed the Command Window, you can reopen it by selecting the `HOME` tab in the toolstrip and finding the group of commands labelled `ENVIRONMENT`. Use the mouse to select the tab entitled `Layout` – this will reveal a pull-down menu. From here choose `Default` from the layout options.

2.2 MATLAB® as a Calculator

At its most basic, MATLAB® can be used as an elaborate scientific calculator – and MATLAB® Mobile™, which can be downloaded on to your phone, is particularly useful for this. There are many advantages to using MATLAB over a standard calculator for more complex and lengthy calculations: in MATLAB you can see your calculation in full, making it easier to spot mistakes, and you can recall and edit previous calculations at any time during your session and even in subsequent MATLAB sessions.

There is one very important difference between a standard calculator and MATLAB – MATLAB was originally designed as a linear algebra tool so the default for multiplying arrays of data is to use matrix multiplication. MATLAB is also able to perform element-by-element multiplication, but there is a subtle, but important, difference in the commands for multiplication, division and powers which tell MATLAB which form of multiplication you require (see ◻ Table 2.1).

If you do not know what the difference between array (element-wise) operations and matrix operations is, don't worry, we will explain them in ▶ Chap. 3. However, it is important to be aware that they are mathematically distinct so that you get into the habit of using the correct notation for multiplication, division and powers. If you do not know anything about matrix algebra, it is likely that you will want to use array operators and so should get used to using the period sign before power, multiplication and division symbols.

To use the command line in the `Command Window`, type in your calculation at the command prompt, and then press ENTER on your keyboard to see the answer. Try this out – there are a few examples below:

Table 2.1 Comparison of the notation for mathematical operators when you want to perform matrix algebra (matrix operators) and when you want to perform element-by-element calculations (array operators). The operations are listed in the order sometimes known by the acronym BODMAS (or BIDMAS or PEDMAS) which is commonly used as a reminder for the order in which operations are performed in calculations which combine more than one operator

Mathematical operator	Matrix operators (Linear algebra)	Array operators (Element-by-element)
Brackets	( )	( )
Order	^	.^
Division	/	./
Multiplication	*	.*
Addition	+	+
Subtraction	-	-

```
5 + 4
4 .* 5
80 ./ 4
3 - 6 .* 2
4.^2
```

Now repeat the above commands, but this time don't include the `dot` before the multiplication and division signs. You should get the same answers. For calculations that do not involve vectors or arrays of data, the inclusion of the `dot` has no effect on the calculation. Now compare the following commands:

```
[1 2 3].*[4 5 6]
[1 2 3]*[4 5 6]
```

The second line of commands should give an error message as we have tried to apply a matrix operation incorrectly – we will examine this in more detail in ► Chap. 3.

The above examples are easy to work out by hand (apart from the last command in which we got an error) – this is deliberate: whenever you are trying out a new feature of MATLAB, conducting data analysis or developing a mathematical representation of a biological system, it is helpful to start by testing your code with a simple example that you know the exact or at least approximate answer to. This provides a check that the code is working as intended.

2.3 Parentheses

Parentheses, or brackets, serve a variety of functions in MATLAB. There are three types of parenthesis which, in general, are not interchangeable (see Table 2.2).

Table 2.2 Summary of the different types and uses of brackets in MATLAB®. Examples will be given as you work through the book. Examples can also be found in the MATLAB documentation by searching for "Operators and Special Characters" (for details on how to access help, see ► Chap. 5)

Bracket type	Notation	Uses in MATLAB®
Round	( )	1. In algebraic calculations to assign precedence of operations 2. To provide arguments for functions 3. To identify the value at any position in a matrix or array (indexing)
Square	[ ]	1. To create arrays, vectors and matrices 2. To join two arrays together (concatenation) 3. To define an empty matrix 4. To delete a column or a row from a matrix 5. To obtain multiple outputs from certain functions
Curly	{ }	1. To construct a cell array 2. To access the contents of a cell within a cell array

To group terms which are multiplied by another term (or a set of terms), use round brackets. For single values, square brackets will work, but they are not strictly correct and will produce a warning message in MATLAB if used in a script (we will learn more about scripts in ► Chap. 4). You must also explicitly include a multiplication sign. Type the following in the Command Window:

```
3.*(6 + 4)
```

Type in the above line of code again, but this time don't include the multiplication symbol (.*). MATLAB should produce an error message. The default colour used to display error messages in MATLAB is red. If you have used other computer languages or packages, you may have become conditioned to think that error messages provide little in the way of information for the non-expert user as they can be difficult to understand. A great advantage of MATLAB is that the developers have worked hard to produce helpful error messages, and not only does MATLAB suggest the most likely cause for your code to fail but also frequently suggests how you could correct your code.

When you tried the above example without the multiplication symbol, you should have received an error message that looks something like

```
3(6+4)
↑
Error: Invalid expression. When calling a function or indexing
a variable, use parentheses. Otherwise, check for mismatched
delimiters.
Did you mean:
>> 3*(6 + 4)
```

Notice that MATLAB does not suggest you precede the asterisk with a `dot` – at this stage the dot is an optional extra that we are including to get into the habit of using. If you are using an older version of MATLAB, you may only get the error message and not the suggested correction; newer versions may also have a differently worded error message.

2.4 Command History

The up and down arrow keys on your keyboard can be used in the `Command Window` to select, edit and rerun previous commands. Another way to see your previous commands is to open the `Command History` window. This is not visible in the standard layout of MATLAB. To find the Command History, navigate to the `HOME` tab in the toolstrip and click on `Layout` to reveal a pull-down menu. From this menu, select the `Command History` menu and choose `Docked` from the list of options. You should now have a new window in your MATLAB interface entitled `Command History`. In this window, you will find all the commands you have entered so far in MATLAB. If you want to rerun a command, select and drag it into the `Command Window` and press ENTER. If you want to rearrange your MATLAB interface, you can drag and drop the windows, change their sizes or choose a standard layout from the `Layout` menu on the toolstrip. Any window can be closed by selecting `Close` from the pull-down menu in the top right-hand corner of that window.

You may see that some of the commands in the Command History have a small red bar next to them. A red bar indicates that there was a fatal error in the command which meant that MATLAB could not evaluate the command. Hovering over the red bar displays an error message explaining why that line of code didn't work.

You should see a red bar next to the last command you wrote in ▶ Sect. 2.3.

2.5 Adding Comments to Lines of Code

When writing commands, it is good practice to add comments to your code so that when you come back to your work another time, or share your code with a colleague, it is straightforward to understand your code and what it does. Any comments in your code are ignored by MATLAB. This will be especially useful when you learn how to write scripts in ▶ Chap. 4. These comments can also be included when writing in the Command Window which can be helpful when going back through the commands you have previously written.

To indicate that you want MATLAB to ignore a line of text, start by typing a % sign. The % sign and any writing after it will automatically turn green to show you that it is a comment and will be ignored by MATLAB. Try out the following in the `Command Window`:

```
4.2+5.1+5.8+2.9  % sum of lengths of 4 male saltwater crocodiles
(4.2+5.1+5.8+2.9 )./4  % mean length
```

2.6 Numerical Constants, Trigonometric Functions, Logs and Exponents

A key component of MATLAB is the vast array of built-in functions included in the package. We have left a detailed explanation of function syntax until ▶ Chap. 6 as it is useful to first get an intuitive grasp of functions, and so we introduce new functions as we progress through the book. MATLAB contains all the usual functions that you can find on a scientific calculator (and many more). The mathematical constant π is a built-in function that takes no input arguments[1]:

```
pi
pi.*3.^2 % calculate the area of circle of radius 3
```

There are separate functions for the basic trigonometric functions depending on whether you want to provide the angle in radians or degrees.

```
sin(pi) % calculate sine of pi in radians
sind(180) % calculate sine of 180 in degrees
```

Notice that the answer MATLAB provides to the calculation `sin(pi)` is not exactly 0. This is because numerical values, including the value of π, are stored as floating-point numbers, so sometimes there is a small rounding error in the result.

Another example of a floating-point error is when we use fractions which cannot be exactly expressed as floating-point numbers:

```
1 - 2./3 - 1./3
```

It is important that you are aware floating-point errors exist and that values which are mathematically zero may not give a computational zero. It does not matter if you don't understand why they exist, although there are plenty of explanations online if you are interested and MATLAB includes a detailed description of floating-point numbers which can be found by typing

```
doc Floating-Point Numbers
```

1 From now on we will not always say "Try the following in the Command Window" instead assume, unless otherwise stated, that you should try any bits of code in MATLAB® as you work through the book.

If you are comparing observed data with calculated values, it is often advisable to include a degree of tolerance in what to accept as an exact match (see the section on Logical Operators in ► Chap. 11).

Exponential and logarithmic values are coded in a similar way to trigonometric functions. Note that in contrast to the language often used in schools, the natural log (base e) is called using the command **log**, and log base 10 is called using the command **log10**. Euler's number, $e \approx 2.7183$, is a function, **exp**, where the input is the power of e required.

```
exp(1)
exp(2)
```

The natural log of Euler's number is:

```
log(exp(1))
```

Compare this to calculations from the next two lines:

```
log10(exp(1))
log10(10)
```

2.7 Assigning Calculation Results to a Named Variable

The solutions to the calculations we have run so far in the Command Window are presented in the form **ans =**

Look in the **Workspace** browser which is part of the MATLAB interface. If you cannot see a window labelled 'Workspace', navigate to the **HOME** tab in the toolstrip, find the **Layout** pull-down menu in a group of commands called **ENVIRONMENT** and select **Default** option from the pull-down list.

If you have implemented any of the commands so far, then the **workspace** should contain a variable named **ans** with a numerical value. The value of **ans** will depend on which calculation you carried out most recently. Now type in the Command Window:

```
7.*8
```

Press ENTER on your keyboard and the value of **ans** should change to 56.

The variable **ans** can now be used in another calculation:

```
ans./2
```

MATLAB has used the current value of **ans**, divided it by two and then assigned the result to the variable **ans** (abbreviation of answer). In general, it is preferable to create your own variable names, since the value of **ans** will be replaced every time you carry out a new calculation. Type the following into MATLAB and after each command look at the contents of the Workspace browser.

```
convertKnots = 0.5144444
windSpeedKnots = 4
windSpeedMperS = windSpeedKnots.*convertKnots
```

This calculation converts wind speed in knots to wind speed in meters per second. The first two lines of commands created two variables which were then used in the third calculation.

Suppose we now want to convert another wind speed to meters per second, say 8 knots, with minimal typing. There are various approaches we can use.

▪ Option 1

Make sure you are in the Command Window by clicking anywhere in it. Now press the up-arrow key on your keyboard. This will enable you to navigate back to the command **windSpeedKnots=4.** In recent versions of MATLAB, pressing the up arrow on your keyboard at the command prompt may open a pop-up window with earlier commands listed in order of execution which you can navigate up and down to select any previous command using the mouse. Once you have located the required command, use the backspace to delete the 4. This should take you out of the list of previous commands, and you can now replace it with an 8:

```
windSpeedKnots = 8
```

Press ENTER and check that **windSpeedKnots** has updated the value in the workspace. Now use the same technique to navigate to the equation to convert the wind speed to meters per second. This time you don't need to edit the command, so, once you have found it, press ENTER to rerun the calculation for the new wind speed.

```
windSpeedMperS = windSpeedKnots.*convertKnots
```

The left-hand variable, **windSpeedMperS**, is updated to a new value in the workspace since the value of **windSpeedKnots** was increased.

▪ Option 2

Double-click on the **windSpeedKnots** variable in the Workspace browser. This should open a new window which contains a spreadsheet. Edit the value in the spreadsheet and press ENTER. You should see the value of **windSpeedKnots** change in the workspace. Now use the up-arrow key on your keyboard to find and rerun the conversion equation as per Option 1.

In MATLAB there are often several equally valid approaches to performing a task. For example, two further options for editing a value are (i) edit the value of

windSpeedKnots by clicking on its value in the Workspace browser and editing it without needing to open the spreadsheet, and (ii) right-click on the variable name in the workspace and select '**Edit Value**' from the menu that appears.

As you get more confident with the MATLAB language and start to use it for your own work, you will probably want to minimise the use of the point-and-click method and instead try to do everything using typed MATLAB commands as it is easier this way to make a record of changes you have made.

2.7.1 Rules for Naming Variables

There are four basic rules to be aware of when you name a variable. Variable names:

- Are case sensitive: i.e. b is not the same as B.
- Must start with a letter: after the first letter, they may include any combination of letters, digits and underscores.
- Must not include spaces.
- Cannot be 1 of the 20 MATLAB keywords such as **for** and **end**. To find a list of keywords, type **iskeyword** into the Command Window and press ENTER.

Variable names can be any length, but MATLAB will ignore any characters beyond the value which can be found by evaluating the command:

```
namelengthmax
```

It is good practice to give your variable a name that describes the information it represents:

```
a = 100 % the name tells me nothing about the value 100
gallWaspFecundity = 100 % attaches meaning to the variable name
```

This is an example of camelCase whereby each word, except the first, starts with a capital letter which makes the variable name easier to read.

Warning
The program will allow you to name a variable the same name as an existing function in MATLAB. The newly defined value of that variable will take precedence over the built-in value.

Consider the following changes to built-in **pi** when it is replaced by a value of 250:

```
pi % built-in value of pi should output 3.1416
pi = 250 % user assigns 250 to the value pi
pi % should now give a value of 250
clear pi % clears the user assigned value of pi
pi % built-in value of pi
```

To check whether a proposed variable name has already been used in your workspace or as a function name, use the command **exist.** The command will return a 0 output if the name given does not exist within MATLAB else it will return a 1 or higher (depending on whether the name has already been assigned to a variable, file, folder and so on).

```
exist checkname % checkname does not exist
exist pi % check pi is a built-in function
exist gallWasp % check gallWasp does not exist
gallWasp = 10
exist gallWasp % gallWasp does exist
```

Tip

If you make a mistake while typing into the command line and you want to start again, pressing CRTL+C will cancel any command that you have started.

2.8 Saving Your Work

2.8.1 Saving all the Commands and Output Which Have Appeared in the Command Window

In general, as you start to use MATLAB in your own work, you should group related commands in script files (we will find out more about how to use script files in ▶ Chap. 4). However, sometimes when you are at a more exploratory stage of work, you may want to keep a record of what you have tried out. For these cases, you may want to make use of the **diary** command.

The **diary** function creates a verbatim copy (excluding graphics) of your MATLAB session in a file from when you activate it by providing an output file name, to when you stop it by using the **diary off** command. Try the following example:

```
diary MyRamblings.out
gallWasp=100
Area=pi.*5.^2
diary off
```

Now open the file MyRambling.out in a standard text editor. If you can't find where MATLAB has stored your file, type

```
pwd
```

into the **Command Window**. This will show the current working directory. The command stands for 'print working directory'. This is where the diary will be saved. If you want to change where the diary is saved, you can use the command **cd** to navigate to the desired location[2], e.g.

```
cd('C:\Users\MyName\MyMatlabWork')
```

A major disadvantage of using **diary** is that if you wanted to rerun the code you recorded in MATLAB, then you need to edit your output file to strip the file of any output that isn't code.

2.8.2 Saving Variables and Parameters from the Workspace

Some more complex models and data analyses involve algorithms which take a long time to generate results, so you may want to save the results to a file for later use. One way to do this is to save the complete contents of the workspace to a MATLAB MAT-file which has the extension **.mat**. You will learn more methods to save data generated in MATLAB to other file formats in ► Chap. 10.

You can either save the whole workspace or use the mouse to select a subset of variables in the Workspace browser (by holding the CTRL key down on a Windows® machine). Once you have selected the variables of interest, leave the mouse where it is and right-click. A pull-down menu should appear from which you can select the option '**Save As...**'. A new pop-up window will appear suggesting a name, location the file will be saved in and the format it will be saved in. The default format is a **.mat** file. Try this out: save the variable **gallWasp** in the file **gallWasp.mat** accepting the default file location (which should be your current working directory in MATLAB).

2.9 Clearing Variables from the Workspace

When starting a new section of work in MATLAB®, it is often a good idea to clear all the variables in the workspace. One command for this is, which you should run now is:

```
clearvars
```

Alternatively, you can use

```
clear
```

If you now run the following command, you should get an error message:

```
gallWasp
```

2 If you are using macOS or Linux, you may need to use forward slash rather than back slash in filepath names.

However, since we saved this variable to a MAT-file, we can add it back into the **Workspace**. To do this, navigate in the **Current Folder** window to the **gallWasp.mat** file and double-click on the file. The variable **gallWasp** should reappear in your workspace.

Sometimes you may only want to clear a subset of the values in the workspace. To do this follow the command **clear** by the names of the variables you want cleared from the workspace:

```
x = 3
y = x + 4
z = x.*y
clear x z
```

Alternatively, you can clear all variables that start with the same sequence of letters and numbers:

```
a1 = 1
a2 = 45
alpha = 234
b1 = 2
clear a* % clear variables which start with an 'a'
```

Another useful command to tidy up the Command Window is:

```
clc
```

This command clears everything from the Command Window but does not change the contents of the workspace. The impact of this command is purely visual.

Take-Home Message

The Command Window can be used to perform mathematical operations and to explore the MATLAB® syntax. You can assign variable names to calculations and use these values in other calculations. Variables which are currently accessible are shown in the Workspace browser. Variable names are case sensitive, cannot start with a number and must not contain any spaces.

The type of brackets you use is important. You need to use the correct bracket type so that you can assign your calculations to a specific data type. The use of a dot before some mathematical operators (*, / and ^) changes the type of calculation performed – this is explored further in ► Chap. 3.

It is good practice to add comments to your code and to clear the workspace when you start a new piece of work.

Arrays and Matrices

C. R. Webb, M. Domijan, *Introduction to MATLAB® for Biologists*,
Learning Materials in Biosciences, https://doi.org/10.1007/978-3-030-21337-4_3

What You Will Learn in This Chapter

In this chapter, you will learn how to create and work with arrays, vectors and matrices. The chapter introduces the subscript and index methods of accessing data from an array and explores methods to create sequences of values. The built-in functions for creating standard matrices are introduced, and you will explore methods to concatenate and augment existing matrices. You will learn why it is important to distinguish between matrix and array operators and see how matrix algebra can be applied in a biological context.

3.1 Arrays and Matrices

MATLAB® was initially developed to tackle numerical problems in linear algebra, and as a result, it has many functions based around the creation and manipulation of matrices.

A matrix is just a rectangular or square array of values – in the final section of this chapter, we show how matrices can be used to express a set of difference equations which then enables us to use matrix algebra to find long-term solutions for the system. In MATLAB there is no distinction between a matrix and an array. The difference lies in which mathematical operations are appropriate for a particular set of values.

Since the origins of MATLAB are in matrix algebra, the program follows the conventions used in matrix algebra to describe data: the array or matrix size (dimensions) is defined by the number of rows by the number of columns (R × C). The location of any element in the matrix is defined by its row number followed by its column number. The row number is always stated first: you can use the mnemonic 'aRC' to remind yourself of this. The first value in the matrix is in row 1 column 1 and is indexed as 1 (this is an important distinction for readers familiar with other packages where the first value in an array may be indexed as 0).

Values in three-dimensional arrays are accessed by giving the row, column and the page (or sheet) number. Arrays can be created with any number of dimensions. A vector is a type of an array that is one-dimensional. Either it has one row and one or more columns (dimensions 1 × n), or it has one column and one or more rows (n × 1).

3.2 Creating and Manipulating Vectors

We will start by exploring the notation used to create row and column vectors. In the first example, we will create a 1 row by 3 column vector and a 3 row by 1 column vector. The types of brackets you use are important: a standard vector or array of values is always enclosed in square brackets. In the Command Window, type:

```
V1 = [1 3 5]
V2a = [1; 3; 5]
V2b = [1
       3
       5]
```

Notice that you can either use a semi-colon to start a new row of data or you can write the next row of the vector by starting a new line.

Look in the **Workspace** browser for the vectors **V1, V2a** and **V2b**. You should see that **V1** has a comma-separated vector, and in **V2a** and **V2b** the values are separated by semi-colons. This shows that **V1** is a row vector and **V2a** and **V2b** are column vectors. When you type in data, you can include commas between values in a vector instead of, or as well as, spaces if you prefer. It makes no difference to the way the data is stored in MATLAB.

```
V3 = [1,7, 9 ,11]
```

The use of **square** brackets to create the vectors is important. Try the above commands again, but this time use round brackets (). This should generate an error message. Now try using curly brackets {}. This won't generate an error message but has stored your data in a different format called a cell. We will explore cells in ▶ Chap. 10 (Importing and Exporting Data).

Note that when typing the commands, the output given by MATLAB might look different to what you typed in the command line. Contrast the output for the following:

```
V4 = [123.4 123.4]
V5= [12.3 1234.5]
```

The vector **V4** outputs the values to four decimal places, while the vector **V5** is displayed in the command window with the common factor of 1000 (written as 1.0e+03) outside the array and the elements of the array scaled accordingly. You can alter the way that MATLAB® displays output using the **format** function, for example:

```
format bank
V5 = [12.3 1234.5]
```

This will change the format of all output in your session until you change it again. For now, change it back to the default setting:

```
format short
```

For a complete list of the display format options, you can explore the documentation file for the built-in function **format**:

```
doc format
```

3.2.1 Transpose of a Vector

Transposing a vector switches a row vector to a column vector (and conversely a column vector to a row vector) keeping all elements in the same order. For a matrix an element in some row *i* and column *j* which we write as position (i, j) is moved to position (j, i). Using the transpose function provides an efficient way of creating column vectors without having to type lots of semi-colons. As is the case for many tasks in MATLAB, there is more than one way to transpose the vector – below are two methods for you to try. The second method follows conventional notation in matrix algebra.

```
V6 = [1 4 9 8]
transpose(V6) % longhand
V6' % shorthand
```

3.2.2 Sequences

The colon operator can be used to efficiently create vectors with a fixed, user specified, step size. If no step size is supplied, the default step size 1 is used.

```
3:8 % creates a vector of the numbers 3 to 8 using default step size of 1
989:1:1002 % explicitly specify a step size of 1
```

Try the following code in the command line and look at the output:

```
0.5:0.1:1
2:-0.2:1
2:0.2:1
```

A negative step size must be given if you want a descending sequence of values. The last value in the command defines an upper boundary for the sequence (or lower bound for a descending sequence). This value will not be included in the sequence if it is lower than the next possible value in the sequence (or higher for a descending sequence) as illustrated by the following examples:

```
0.5:1:4.1
100:7:67
100:-7:67
V7=[0.87:0.34:3]'
```

Two or more sequences can be combined in one vector, but this time you need to include square brackets:

```
[1:1:5 99:-3:90 101:150:500]
```

3.3 Suppressing Output

Sometimes we do not need to see the output from a command in the Command Window. Inserting a semi-colon at the end of a line of code will stop the output generated from that line from being displayed in the Command Window. Allowing all outputs to be written to the Command Window can increase the run time of your programs, so when you start to write scripts (in ► Chap. 4), you will see that if you do not include a semi-colon at the end of a line of code, MATLAB generates a warning suggesting that you add a semi-colon.

```
1:7:48;
```

3.4 Extracting Subsets of a Vector

There are two ways of defining the location of an element in a vector:

3.4.1 Method 1: Linear Index

Each value in the vector has an associated linear index. For a vector with n elements, the linear index counts in steps of 1 from the first element in the vector (first row for a column vector and first column for a row vector) which is in position 1, up to the last value which is position n (this is different to some other computer languages which index from 0 to n-1). To extract values from a vector, provide the name of the variable followed by the locations to extract in round brackets.

```
CatWeight = [1.3 1.7 1.65 1.4 1.5]; % row vector
CatWeight(1) % extract value in position 1
CatWeight(3) % extract value in position 3
CatAgeWeeks = transpose([12 15 13 10 14]); % column vector
CatAgeWeeks(2)
CatAgeWeeks(5)
```

Groups of values can be extracted by specifying a list of locations using either a sequence or a vector of locations:

```
CatWeight(1:2:5) % Cat weights in position 1, 3 and 5
CatAgeWeeks([1 2 4]) % Cat ages in positions 1, 2 and 4
```

3.4.2 Method 2: Subscript

The subscript method locates an element using the row number of the element followed by the column number.

```
CatAgeWeeks(1,1)
CatAgeWeeks(3,1) % the value in the 3rd row and 1st column
```


Try the command:

```
CatAgeWeeks(1,3)
```

This will produce an error as third column does not exist in the **CatAgeWeeks** variable.

3.5 Augmenting a Vector

The MATLAB language does not require you to declare types and sizes of variables before you use them. This means that you can increase the size of a vector by allocating a new value to a previously undefined position in the vector:

```
CatWeight(6) = 1.2
CatWeight(9) = 1.7
```

The first command increases the size of the vector by one element and allocates the given value to that element. The second command increases the size of the vector by a further three elements; however, as only a value is given for the element in position 9, the elements in positions 7 and 8 are automatically allocated a zero.

If we call a value that doesn't exist, then MATLAB will produce an error as it has no value stored in the requested position:

```
CatAgeWeeks(6)% this command will produce an error
```

3.6 Mathematical Operators

3.6.1 Multiplication of Vectors

There are two mathematically distinct methods for multiplying one-dimensional arrays (or vectors) of data. Since the MATLAB language was originally designed to solve problems in linear algebra, the default mode of multiplication is matrix multiplication. The alternative mode is element-wise multiplication.

Suppose we want to multiply two vectors, A and B. For both modes of multiplication, the number of elements in vector A and in vector B must be the same. For matrix multiplication (*) A and B can only be multiplied (in that order) if the number of columns in vector A is equal to the number of rows in vector B. For element-wise multiplication (.*),

A and B just have to have the same number of elements and can be any combination of row and column vectors.

For example, create two vectors:

```
PropCropReject = [0.2 0.1 0.3]; %row vector dimensions: 1 row by 3
% columns
TotalYield = [10; 25; 43]; %column vector: 3 rows by 1 column
```

Using element-wise multiplication, we can multiply either vector by itself:

```
PropCropReject.*PropCropReject
TotalYield.*TotalYield
```

However, if you try to multiply two column vectors or two row vectors with notation $*$, you will encounter a MATLAB error message – try it out:

```
PropCropReject*PropCropReject
TotalYield*TotalYield
```

Multiplying a column vector by a row vector by either method will give the same answer:

```
TotalYield.*PropCropReject
TotalYield*PropCropReject
```

The final option, multiplying a row vector by a column vector, is perhaps the most important to be aware of as **`both methods produce an answer, but not the same answer`** (see ◘ Fig. 3.1 for mathematical details), so we may think our program is running fine, but the results will be completely wrong if you have unintentionally used the wrong form of multiplication for your analysis:

```
PropCropReject.*TotalYield
PropCropReject*TotalYield
```

3.6.2 One-Dimensional Arrays: Division

Matrix division is a bit more complicated, and we will not go into details here. Element-wise division works in the same way as element-wise multiplication but replaces the multiplication sign in each cell of the solution with a division sign.

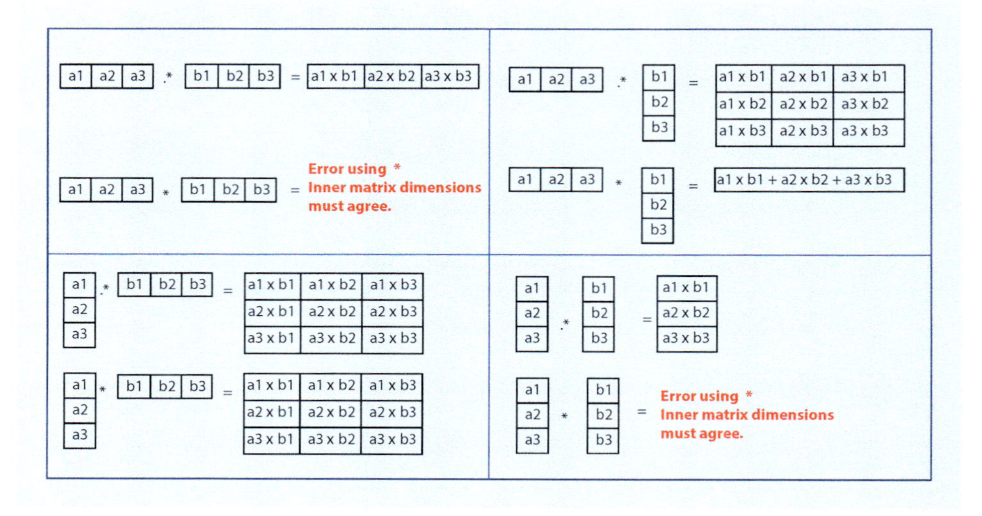

Fig. 3.1 Comparison of element-wise multiplication (using .* in MATLAB®) and matrix multiplication (using * in MATLAB®)

3.6.3 One-Dimensional Arrays: Addition and Subtraction

There is no difference between element-wise addition and subtraction and matrix addition and subtraction. The rules are the same as for element-wise multiplication, replacing the multiplication sign with a plus or a minus sign as required.

```
PropCropReject + PropCropReject
TotalYield - TotalYield
PropCropReject + TotalYield
```

3.7 Creating and Manipulating Matrices

The term matrix is used in this book to refer to any square or rectangular array of data. The main exception to this is when we discuss matrix multiplication in which case we mean multiplication as defined in linear algebra as opposed to element-wise multiplication. A vector is a special case of a matrix with either one row or one column. As with vectors, there is more than one way of entering matrix data into MATLAB:

- Start with a left square bracket. [
- Separate values in the same row of data using either a space or a comma.
- Start a new row of data using either a semi-colon or by pressing ENTER to start a new line.
- End the matrix with a right square bracket].
- Use a semi-colon after the second bracket if you want to suppress the output to the screen.

- Every row must have the same number of data points.

```
PatientHB = [9 14 12 10; 8 7 6 10; 12 15 13 5; 11 12 32 12];
NumberSnails = [10 12 34
        23 34 9
        67 1 43
        9 12 29];
```

- You can also enter the values in the Variables window by creating an empty matrix and then opening the variable as follows:

```
EmptyMatrix = [];
open EmptyMatrix
```

Try out this method: suppose your data consists of lots of zeros and a few non-zero values – you can just place non-zero values in the desired location in the array. Any elements you do not fill in will automatically be filled with a zero to ensure all the columns are the same length and all the rows are the same length. Importing data will be covered in ► Chap. 10.

3.7.1 Adding Data to an Existing Matrix (Concatenating)

To add a new row of data to an existing matrix, include the original matrix inside square brackets, and use a semi-colon to start a new line of data:

```
PatientHBNew = [PatientHB; 17 19 8 12]
```

To add a new column, first create a vector including the new column of data. Now you can join the two sets of data together:

```
ExtraWardData = [14;12;7;17;12];
AllHBData = [PatientHBNew ExtraWardData]
```

Alternatively, you can include the new column without having to define it first by including it as an array which concatenates to the original array:

```
A = [23 23; 89 23]
ANew = [A [1;5]]
```

There are a variety of built-in functions which can speed up the creation of matrices which contain repeating patterns. Suppose, for example, you want to create a 4 by 3 matrix in which every column has the numbers 1 to 4. While you could type all the values in, or take a shortcut by typing in sequences, MATLAB provides an even quicker way of doing this via the built-in function **repmat** (short for 'replicate matrix') so that you only type the sequence once and then use the command to duplicate the vector:


```
Column1 = [1:4]';
ThreeColsTheSame = repmat(Column1,1,3)
DoubleLengthCol = repmat(Column1,2,1)
```

3.7.2 Matrix Dimensions

The size of all active variables is displayed in the Workspace browser. There are occasions however when it is useful to create parameters which describe the size of the data set. The size of a matrix can be defined either in terms of the number of rows and columns or as the total number of elements.

```
MyBigData = repmat([1:4:98],3,5);
size(MyBigData) % prints number of rows and columns to the screen
[nrowsMBD, ncolsMBD] = size(MyBigData) %stores the number of rows and
% columns to user defined variables
NumElementsMBD = numel(MyBigData) %counts the total number of
% elements in the matrix
```

MATLAB also includes a function called **length** – be careful with this command – it returns the length of whichever is larger of the number of rows and the number of columns, so it may not give the answer you are expecting if you are writing code to use on several data sets.

```
LongestDimension = length(MyBigData) % be careful with this command
```

3.7.3 Reshape and Transpose a Matrix

Reshape allows you to adjust the layout of the elements from a matrix or vector into one with the same total number of elements but a different shape. One useful application of this is if we wanted to create a 10 by 10 grid of the numbers 1 to 100. Rather than concatenating ten vectors, we can create a vector of the required numbers and then use reshape to get them to the required format:

```
PatientID = [1:100];
PatientIDGrid = reshape(PatientID,10,10) % numbers going down columns
PatientGridv2= transpose(PatientIDGrid) % numbers going across rows
```

Another example:

```
MyMatrix = 1:16;
OddRowEvenRow = reshape(MyMatrix,2,8);
OddColEvenCol = OddRowEvenRow'% transpose
```

Some of the above examples may initially seem rather specialized; however, as you start to deal with larger arrays of data, these shortcuts can save a lot of time and effort.

3.7.4 Extracting Values from a Matrix

Every location in an array can be extracted using either the index or the subscript method. Indices start from a value of one in the top left-hand corner of the matrix and count down the columns (◘ Fig. 3.2). Subscripts identify elements by their row and column numbers. Which method you use is in part a matter of personal preference and in part context driven. The main advantage of using subscripts is that if you add or delete rows, the subscript location of the remaining cells will stay the same. In contrast, the index of values in the second column onwards will change to reflect the new number of rows (◘ Fig. 3.2).

To explore the impact of adding a row to data on the extracted value, try the following code:

```
DryWeight = [23 45 89; 43 56 52; 94 34 50];
DryWeight(6) % find the value in index position 6
DryWeight(3,2) % find the value in row 3 column 2
DryWeight = [DryWeight; 40 23 12]; % add another row of data
DryWeight(6) % find the value in index position 6
DryWeight(3,2) % find the value in row 3 column 2
```

A

Indices				Subscripts			
1	6	11	16	1,1	1,2	1,3	1,4
2	7	12	17	2,1	2,2	2,3	2,4
3	8	13	18	3,1	3,2	3,3	3,4
4	9	14	19	4,1	4,2	4,3	4,4
5	10	15	20	5,1	5,2	5,3	5,4

B

Indices			Subscripts		
1	5	9	1,1	1,2	1,3
2	6	10	2,1	2,2	2,3
3	7	11	3,1	3,2	3,3
4	8	12	4,1	4,2	4,3

◘ **Fig. 3.2** Comparison of the two methods of accessing an element in an array in two examples, A and B. Removing a row from the first matrix results in all values from the start of the second column onwards being relabeled for the index method of accessing the array, whereas the names remain the same in the subscript method

The colon operator, : , works in different ways depending on whether we use it as part of the index or subscript method and also still works to create a list of values:

■■ Index method

```
ListDryWeight = DryWeight(:) % list values by index order as column
% vector
RowDryWeight = DryWeight(1:numel(DryWeight)) % lists values by index
% order into a row vector
SomeDryWeights = DryWeight(3:10) % extracts dry weight in positions 3
% to 10 inclusive.
EveryOtherWeight = DryWeight(1:2:numel(DryWeight)) %extracts every
% other value
SpecificDryWeights=DryWeight([1 7 11]) % extracts the dry weights in
% positions 1,7 and 11
```

■■ Subscript method:

```
FirstRow = DryWeight(1,:) % extracts first row and all columns
ThirdCol = DryWeight(:,3) % extracts values for all rows for the 3rd
% column
TwoByTwo = DryWeight(1:2,1:2) % extracts square of data from top left
% hand corner
```

3.7.5 Built-in Matrices

In this section, we introduce some common MATLAB functions which provide the basic building blocks for the creation and manipulation of matrices.

The function **`zeros`** creates a matrix of a user-defined size where every element in the matrix is zero. One use of the zero matrix is to provide a matrix in which to fill in values as they are calculated: this is more computationally efficient than adding rows and columns as we go along.

This function can either take one or two inputs. Providing a single input will create a square matrix of the specified size. Providing two inputs will give a matrix which the first number giving the number of rows in the matrix and the second the number of columns.

```
zeros(3)
zeros(2,4)
```

The **`ones`** function creates a matrix with user-specified dimensions in which every element has the value 1.

```
ones(2)
ones(4,5)
```

The square identity matrix is the matrix algebra equivalent to the number 1[1].

```
eye(4)% this is the identity matrix
eye(3,5) % this a the matrix with ones along the main diagonal
eye(5,2) % this a the matrix with ones along the main diagonal
```

The built-in functions **randi** and **rand** can be used to generate a matrix of pseudorandom integers or values in the range (0,1), respectively.

```
randi(89,4,5) % 4 by 5 matrix of random integers in the range [1,89]
randi(67,6) % 6 by 6 matrix of random integers in the range [1,67]
rand(3,4) % generates a 3 by 4 matrix of random numbers in the range (0,1)
```

Each time you execute these commands, you will get a different answer (remember you can use the up arrow to recall a command to save retyping). If you want to create a repeatable result, then you can fix the 'seed'. The seed is a numerical value used to initialize a pseudorandom number generator. Try the commands below to find out how to fix the seed, and compare the results for the three matrices generated below:

```
s = rng; % store the current value of the seed to a variable 's'
M1 = rand(2) % generate a 2 by 2 matrix of numbers in the range (0,1)
M2 = rand(2) % generate another 2 by 2 matrix of numbers in the range (0,1)
rng(s) % resets seed to value at the start of this section of code
M3 = rand(2)
```

3.8 Comparison of Matrix and Element-by-Element Multiplication

Try out the following commands to compare element-by-element multiplication and matrix multiplication by a matrix of zeros, a matrix of ones and the identity matrix. Do you get the same answer if you use the operator .* or * ?

```
Trial = randi(10,4);
Trial.*zeros(4)
Trial*zeros(4)
Trial.*ones(4)
Trial*ones(4)
Trial.*eye(4)
Trial*eye(4)
```

1 Descriptive text is included after some commands as comments as we found this neater than trying to include it in the paragraphs. Anything after a % sign is not read by MATLAB. You do not need to type this information into the Command Window.

It is important that you choose the correct operator. Suppose we want to use MATLAB to calculate the BMI of a random selection of 100 adults from each of 7 major UK cities. Our input data are two 100 by 7 arrays of data. The first array contains the height in centimeters of each adult in the study, and the second array contains the weight in kg of each adult in the study. We can import this data into MATLAB as two matrices H and W say (we will see how to do this in ▶ Chap. 10).

To calculate the BMI of each adult, we need to use the following formula:

$$BMI = \frac{W}{(H \times 0.01)^2}$$

In MATLAB, we can achieve this in one short line of code (rather than getting you to type in two 100 by 7 tables of data, we have added two lines of code with a subset of the data, where three individuals in two of the cities are identified):

```
W = [75 70 73;65 60 55];
H = [157 180 183;123 144 162];
BMI = W./((H.*0.01).^2)
```

Notice we have used a dot '.' before the multiplication, divide and power signs. This tells MATLAB we want it to perform element-by-element manipulation of the data **not** linear algebra.

Type the above code in – what happens if you miss out the period before the / sign, the * sign or the ^ sign (try them out one at a time). Sometimes leaving the period off has no impact, for example, when we multiply height by a scalar, however, the period ensures that the operations are applied element-by-element, and so it is safest just to include it everywhere.

Examples of element-by-element multiplication (array operator) of two matrices are given in ◘ Fig. 3.3. Element-by-element multiplication will only work on matrices that have the same dimensions.

Matrix multiplication is not the same as element-by element multiplication. Two matrices, X and Y say, can only be multiplied as X×Y if the number of columns in X is equal to the number of rows in Y (i.e. the inner dimension match). The resultant matrix will have the same number of rows as X and same number of columns as Y. Two examples are shown in ◘ Fig. 3.4. Compare the results for multiplying two 2 by 2 matrices with those given in ◘ Fig. 3.3. Some square (i.e. number of rows = number of columns) matrices are invertible such that XY = YX. However, this is not true for all matrices.

$$\begin{pmatrix} x_{11} & x_{12} \\ x_{21} & x_{22} \end{pmatrix} .* \begin{pmatrix} y_{11} & y_{12} \\ y_{21} & y_{22} \end{pmatrix} = \begin{pmatrix} x_{11}y_{11} & x_{12}y_{12} \\ x_{21}y_{21} & x_{22}y_{22} \end{pmatrix}$$

$$\begin{pmatrix} x_{11} & x_{12} & x_{13} \\ x_{21} & x_{22} & x_{23} \end{pmatrix} .* \begin{pmatrix} y_{11} & y_{12} & y_{13} \\ y_{21} & y_{22} & y_{23} \end{pmatrix} = \begin{pmatrix} x_{11}y_{11} & x_{12}y_{12} & x_{13}y_{13} \\ x_{21}y_{21} & x_{22}y_{22} & x_{23}y_{23} \end{pmatrix}$$

◘ **Fig. 3.3** Array (element-by-element) multiplication of matrices X and Y. Calculation of the element in the first row and first column is highlighted in purple

Fig. 3.4 Matrix multiplication of matrices X and Y. Calculation of the first element of X×Y is highlighted in purple

$$\begin{pmatrix} x_{11} & x_{12} \\ x_{21} & x_{22} \end{pmatrix} * \begin{pmatrix} y_{11} & y_{12} \\ y_{21} & y_{22} \end{pmatrix} = \begin{pmatrix} x_{11}y_{11}+x_{12}y_{21} & x_{11}y_{12}+x_{12}y_{22} \\ x_{21}y_{11}+x_{22}y_{21} & x_{21}y_{12}+x_{22}y_{22} \end{pmatrix}$$

$$\begin{pmatrix} x_{11} & x_{12} \\ x_{21} & x_{22} \end{pmatrix} * \begin{pmatrix} y_{11} \\ y_{21} \end{pmatrix} = \begin{pmatrix} x_{11}y_{11}+x_{12}y_{21} \\ x_{21}y_{11}+x_{22}y_{21} \end{pmatrix}$$

There is no difference between scalar multiplication using either type of operator as you can see by comparing the output from the following:

```
A = randi(5,2,3)
2*A
A*2
2.*A
A.*2
```

Matrix division (right and left division) is less straightforward. Further details on matrix operators are provided in the MATLAB documentation site which you can access by typing in the Command Window:

```
doc array vs matrix operations
```

3.9 Application of Matrix Algebra to Biological Systems

Matrix algebra has many practical applications. One application of interest to natural scientists is the use of matrix algebra to investigate population dynamics using discrete time Markov Chain models. This type of model is used to explore systems in which processes can be divided into discrete time intervals such as systems where there is clear seasonality.

3.9.1 Modelling Population Dynamics Using Matrix Algebra: Discrete Time Markov Chains

A nesting box is placed in a tree in the back garden, and the owner wants to know what they are likely to get in the box. There are three possible annual scenarios: the box is unoccupied (E), the box is occupied but no chicks are produced (A) and the box is occupied and chicks are produced (C). Long-term records of nesting boxes on a nearby nature reserve suggest the single yearly transition probabilities sketched in Fig. 3.5.

For example, if the nest was empty this year, then there is a probability of 0.6 that it will be empty next year, a probability of 0.3 that it will be occupied but no chicks are produced

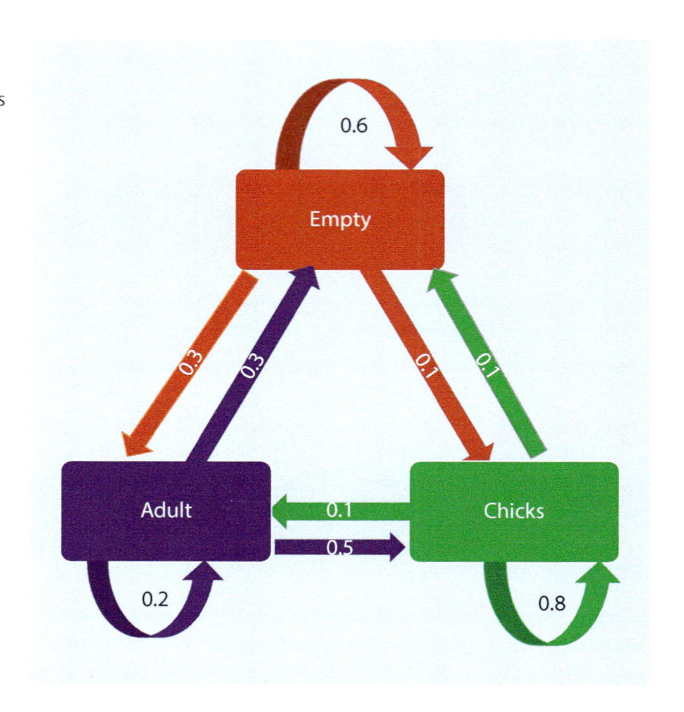

Fig. 3.5 Diagram of transition probabilities between boxes that are empty, with chicks and only with adults

and a probability of 0.1 that it will be occupied and chicks will be produced. We can write this information down as a set of discrete time equations:

$$E_{i+1} = 0.6E_i + 0.3A_i + 0.1C_i$$
$$A_{i+1} = 0.3E_i + 0.2A_i + 0.1C_i$$
$$C_{i+1} = 0.1E_i + 0.5A_i + 0.8C_i$$

An alternative and convenient way to write down these equations is in the form $\mathbf{x}_{i+1} = \mathbf{A}\mathbf{x}_i$:

$$\begin{pmatrix} E_{i+1} \\ A_{i+1} \\ C_{i+1} \end{pmatrix} = \begin{pmatrix} 0.6 & 0.3 & 0.1 \\ 0.3 & 0.2 & 0.1 \\ 0.1 & 0.5 & 0.8 \end{pmatrix} \begin{pmatrix} E_i \\ A_i \\ C_i \end{pmatrix}$$

The box is initially empty when placed in the garden:

$$\begin{pmatrix} E_0 \\ A_0 \\ C_0 \end{pmatrix} = \begin{pmatrix} 1 \\ 0 \\ 0 \end{pmatrix}$$

The probability of having chicks in the first year is therefore $\mathbf{x}_1 = \mathbf{A}\mathbf{x}_0$:

$$\begin{pmatrix} E_1 \\ A_1 \\ C_1 \end{pmatrix} = \begin{pmatrix} 0.6 & 0.3 & 0.1 \\ 0.3 & 0.2 & 0.1 \\ 0.1 & 0.5 & 0.8 \end{pmatrix} \begin{pmatrix} 1 \\ 0 \\ 0 \end{pmatrix} = \begin{pmatrix} 0.6 \\ 0.3 \\ 0.1 \end{pmatrix}$$

Suppose we want to find the probability, we have chicks in 5 years' time. The probability distribution for year 2 is $\mathbf{x}_2 = \mathbf{A}\mathbf{x}_1 = \mathbf{A}(\mathbf{A}\mathbf{x}_0) = \mathbf{A}^2\mathbf{x}_0$, and so by induction it follows that the probability distribution for year 5 is $\mathbf{x}_5 = \mathbf{A}^5\mathbf{x}_0$. It is easy to find the probability distribution for year 5 using MATLAB:

```
A = [0.6 0.3 0.1; 0.3 0.2 0.1; 0.1 0.5 0.8];
x0 = [1 0 0]';
x5 = (A^5)*x0
```

Now type the following code to plot the long-term probability distribution (you will learn more about plotting and loops as you proceed through the book)

```
for i = 1:20
    x(:,i) = (A^i)*x0;
end
plot(0:20,[x0 x])
xlabel('year')
legend('Empty','Adults','Chicks')
ylabel('probability')
```

The method we used above is rather clunky and inefficient. Another way we can find out the long-term probability of seeing chicks is to use the following formula for every generation m:

$$x_m = A^m x_0 = \alpha(\lambda_1)^m v_1 + \beta(\lambda_2)^m v_2 + \gamma(\lambda_3)^m v_3$$

where parameters λ_1, λ_2 and λ_3 are the so-called eigenvalues of the matrix A and ν_1, ν_2 and ν_3 are vectors that represent the eigenvectors of A corresponding to the eigenvalues. Symbols α, β and γ are additional parameters of the model. We need to find the values of all these parameters.

Do not worry if you have never seen any of this before – the reason we have included it is to highlight some of the matrix algebra features of MATLAB and to demonstrate that good use of matrix algebra can save a lot of computational time.

First, we calculate the eigenvalues and the corresponding eigenvectors using the built-in function **eig**:

```
[V,D] = eig(A)
```


The outputs of the function are two matrices – the first is the matrix **V** which contains the eigenvectors as its columns, and the second is a matrix **D** which contains the eigenvalues along its main diagonal and zero entries elsewhere.

We can separate out the eigenvalues and the eigenvectors:

```
L1 = D(1,1); v1=V(:,1); % first eigenvalue & corresponding eigenvector
L2 = D(2,2); v2=V(:,2); % second eigenvalue & eigenvector
L3 = D(3,3); v3=V(:,3); % third eigenvalue & eigenvector
```

Now we need to find α, β and γ. Using the initial conditions, we can write the preceding equation for the case $m = 1$ as:

$$x_0 = \begin{pmatrix} 1 \\ 0 \\ 0 \end{pmatrix} = \alpha \lambda_1{}^0 v_1 + \beta \lambda_2{}^0 v_2 + \gamma \lambda_3{}^0 v_3 = \alpha v_1 + \beta v_2 + \gamma v_3 = \alpha V(:,1) + \beta V(:,2) + \gamma V(:,3)$$

This is the matrix form of a system of three simultaneous linear equations with unknowns α, β and γ. To solve this in MATLAB, we can supply the matrix **V** and the initial conditions to the built-in function **linsolve** to solve for the α, β and γ values that satisfy the last equation above:

```
Parameters = linsolve(V,x0)
alpha = Parameters(1,1);
beta = Parameters(2,1);
gamma = Parameters(3,1);
```

We now have all the information we need to find the transition probability for any year:

```
year=5; %year of interest
xYear=alpha*L1^year*v1+beta*L2^year*v2+gamma*L3^year*v3
```

Compare the results with those found using the matrix multiplication method – try altering the year number.

Take-Home Message

The terminology arrays, vectors and matrices are important outside MATLAB®; however, within MATLAB they are all represented in the same way. Square brackets are used to start and end arrays of data. The semi-colon can be used within arrays to indicate a new line of data or outside arrays to suppress output to the Command Window.

Matrix operators for multiplication, division and powers are not the same as array operators. Always use array operators (i.e. place a dot before the operator) if you are not using matrix algebra. You will know if you should be using matrix algebra.

To access elements in an array, you can use either the linear index method or the subscript method – it is important you understand both methods. Linear indexes start from 1 in MATLAB. There are many shortcuts available in MATLAB from which you can create a wide range of matrices reducing the risk of data entry errors and saving time.

Writing Script Files

C. R. Webb, M. Domijan, *Introduction to MATLAB® for Biologists*,
Learning Materials in Biosciences, https://doi.org/10.1007/978-3-030-21337-4_4

What You Will Learn in This Chapter

The MATLAB® command line is suitable for checking simple calculations and exploring how individual functions work. However, for most tasks it is more efficient to write your code in script files. In this chapter you will learn how to create, edit and save script files, add descriptive text that automatically generates a documentation file for your script, split your code into manageable sections, use the MATLAB Editor to identify coding errors and use it to help debug code and determine how much each command in your code contributes to the overall run time. The chapter will cover writing scripts in both the standard editor and the Live Editor, a relatively recent feature of MATLAB that enables live scripts to be created which can include formatted text, images, equations and user-controlled input.

4.1 What Is a Script File?

A script file is a text file which contains one or more MATLAB® commands. There are two types of MATLAB script file:

- Standard script file:
  - Original form of script files compatible with all versions of MATLAB.
  - Saved with the extension `.m`.
  - Can be read using a standard text editor such as WordPad.
  - When the script is executed, output which is not suppressed in the script is output in the Command Window, and plots are shown in a separate Figure window.
- Live script file:
  - Introduced in MATLAB version 2016a with significant updates in each subsequent release.
  - Saved with the extension `.mlx`.
  - Can only be read within the MATLAB environment and are not compatible with older versions of MATLAB (pre-2016a).
  - Allows the user to switch between code mode and text mode. Text mode can be used to create user-friendly scripts which contain formatted text, images and equations to produce a document that integrates code with detailed descriptions.
  - Interactive user functionality can be added, such as control bars for parameter values.
  - All output, including plots, are displayed within the script window either in a split window to the right of the code or directly below each section of code.

4.2 Why Use Script Files?

In ▸ Chaps. 2 and 3, we introduced the basic syntax of MATLAB by writing commands directly into the MATLAB Command Window. This allowed us to immediately see the output produced by each command. All commands written in the Command Window are automatically saved in the Command History (see ▸ Chap. 2) from where they can be dragged, using the mouse, back into the Command Window. This method becomes cumbersome if you want to rerun a group of commands with different parameter values or if you have several similar versions of the commands, some of which may contain errors, in your history.

In general, it is more efficient to write commands into a script file which you can edit, add comments to and rerun when needed. Script files enable you to repeat the exact same set of commands at any date in the future and can be shared with other users. We strongly advise that you get used to writing in scripts rather than in the Command Window and for most of the rest of the book we will assume that you will write the code into a script before executing[1] it.

There are many other advantages to writing commands in scripts including colour coding of code, error checking, grouping related commands together and adding detailed explanations so that you can understand why you chose to use a given set of commands when you come back to your code at some later date. In this chapter we will introduce the main features of standard script files. We will then introduce some of the additional features available in live script files.

4.3 Creating a Standard Script File

For the main exploration of scripts in this chapter, we will use standard script files since (i) they are compatible with all versions of MATLAB and, (ii) in most cases, the same code written in a standard script file has a shorter run time than when written in a live script. Whether this speed difference is important depends on how complex your code is and whether the difference is parts of seconds, minutes or hours.

Before you start writing scripts, it is a good idea to plan where on your computer you are going to store them. The next section guides you through this process.

4.3.1 Storing and Accessing Script Files

First set up a folder where you will store your scripts. You can achieve this either by navigating around the Current Folder browser in the MATLAB interface or you can enter instructions into the Command Window. If the Current Folder browser is not visible in your MATLAB interface, you can retrieve it by selecting the **HOME** tab from the toolstrip, then clicking on the **Layout** icon and selecting **Default** from the layout options. Alternatively, you can open the Current Folder browser from the Command Window:

```
filebrowser
```

You can now navigate your way around the Current Folder browser and create new directories and files in a similar way to any interactive file explorer. For the rest of this section, we will focus on the command-driven approach to exploring file structure and creating new directories as it is sometimes useful to be able to create new folder from within a script.

1 Executing code refers to the process of 'running' the code either by calling it from the Command Window or from the MATLAB 'EDITOR' or 'LIVE EDITOR' menus on the toolstrip.

We will start with a few useful commands which show the directory MATLAB is currently in and where MATLAB will look for script files. The current working directory is the first place MATLAB will look for scripts and functions, it is also where scripts will be saved to if you don't specify a file path. To find the current working directory, type:

```
pwd
```


This stands for '**p**'rint '**w**'orking '**d**'irectory, and if you have the Current Folder browser open, the output printed should match the location listed below the toolstrip menu bar.

To display a list of all files and folders in the current directory in the Command Window, use the command **dir**. To refine the output so that you can only see the MATLAB files in the folder, use **what**:

```
dir
what
```

To navigate to the location in which you want to create your new folder, use the command **cd** followed by the location. For example:

```
cd c:\users % choose a suitable location
```

If you are using a macOS, you might need to replace the backslash with a forward slash symbol /. Now create a new folder to store your scripts in:

```
mkdir MyTestScripts
```

If you do not have the Current Folder browser open, use the command **filebrowser** to open it.

Notice that the **MyTestScripts** folder is shaded light grey. This indicates that the folder is not in the MATLAB search path. You can list all the directories in which MATLAB will search for a function or file name using:

```
path
```

You should see a long list of paths that contain the MATLAB files and functions which come with your installation of MATLAB. If the directory containing your script file is not in the current path, then MATLAB returns an error '**undefined function or variable**'. To make the folder visible to MATLAB, you need to add the folder to the search path. From the command line, you can use the **addpath** function, specifying the location of the folder you want MATLAB to search in, for example:

```
location = pwd;
addpath([location, '\MyTestScripts'])
```

Alternatively, you can right-click on the folder you want to add to the path in the Current Folder browser: this brings up a pull-down menu on which you need to select `Add to Path` and then either `Selected Folders` or `Selected Folder and Sub-folders`. The selected folder name will change from grey to black indicating that it is now in the MATLAB search path. This enables you to access any files in this folder from MATLAB even if you are not working in the directory where that file is located.

If you want to find more about path options, you can read the MATLAB documentation by typing in the Command Window prompt:

```
doc File Operations
```

and

```
doc addpath
```

4.3.2 Creating and Saving a New Script File

As with many aspects of MATLAB, there is more than one way of creating a new script. Here we present three options; the first method you choose will add the Editor to your interface – you can have several scripts open at the same time.

- Method 1: Select the `HOME` tab from the toolstrip, and click on the `New Script` icon (note: for now, we are sticking with standard script files so do not select the `New Live Script` icon).
- Method 2: Highlight a command (or group of commands) in the `Command History Window`, right-click and then select the option `Create Script` from the pull-down menu. This will open the Editor and insert the highlighted command(s) into the script.
- Method 3: In the `Command Window,` type `edit`. By default, this opens a new standard script file. If you type edit followed by the name of the file, a window will appear with a prompt to confirm if you want to create a new file by this name (unless a file with this name already exist). Using the extension `.m` will open a standard script file, while using the extension `.mlx` will open a live script file.

Use one of the above methods to create a new script file, and save it as `MyFirstScript.m` (to save the file, select the `EDITOR` tab from the toolstrip, and click on the `Save` icon – make sure that you save it in the directory you created in the last section and that this

Fig. 4.1 The command strip containing basic editor functions is located on the top-right of the main interface window of MATLAB

directory is in the current search path). You do not need to give the file extension as MATLAB will automatically assign a `.m` ending if the script is standard script or a `.mlx` ending if the script is a live script. The file should now be listed in the Current Folder window.

Any edits you make to a script file are only saved when you actively choose to save them. Any files in the Editor with an asterisk next to their name have been edited but not yet saved. A quick way to save, undo or redo commands is to use the icons on the menu which you can find along the top right of the MATLAB main interface (see Fig. 4.1). Hovering with the mouse over any of these symbols will reveal a pop-up message with a short explanation of each symbol.

4.3.3 Choosing a Name for a New Script File

When naming a script, there are several points you should be aware of:

- You must start a file name with a letter; you can then use digits, roman letters and underscores.
- You should not include any spaces in your filenames.
- MATLAB will not prevent you from naming a script the same thing as an existing script or built-in function (providing it is in a different directory). This can cause catastrophic programming errors as the default hierarchy of the MATLAB search path places any folders you have added to the search path above built-in functions and variables. To avoid this, you can check whether MATLAB already uses your proposed filename elsewhere using the command **exist** and, if it is already in use, where it has been used using the command **which**:

```
exist MyFirstScript
which MyFirstScript
```

The command **`exist`** returns a 0 if this name does not exist and returns a 1 or higher if this name is taken by another variable in the workspace, another file or another folder.

4.3.3.1 Suggested MATLAB® Documentation Files

If you want to know more about how to find out whether a name of a file already exists and which type of file it is, read the MATLAB documentation files which can be found by typing the following into the Command Window:

```
doc exist
doc isfile
doc isfolder
```

4.4 Editing a Script File

If you have not already done so, create a new empty script file, and save it as `MyFirstScript.m`.

4.4.1 Adding Explanatory Text to a Script File

It is generally a good idea to add some explanation of the content of your script file at the start of the script. Edit `MyFirstScript.m` so that it contains the following text (you can write whatever you like, but you must include the % signs at the start of each line).

```
% MyFirstScript
% Plots a Michaelis-Menten curve
% My name
% Created: today
```

Save your script and now type into the Command Window:

```
doc MyFirstScript
```

This should bring up help documentation which includes the text we added to **MyFirstScript** and a link to the script file. Only comments added to the beginning of the script file will be included in the help documentation.

4.4.2 Adding Commands to a Script File

Next, we will add some commands to `MyFirstScript.m` to:
- Clear all existing variables from the workspace.
- Close any figure windows.
- Define the range of the independent variable.
- Define two parameter values.
- Evaluate the Michaelis-Menten function for these values.
- Plot the function.

It is often a good idea to start a script with commands to clear existing variables from the workspace and close all existing figure windows. This means that any variables left over from previous work won't get accidentally used in your new script. Add the code below into `MyFirstScript.m`:

```
clear
close all
```

Save **MyFirstScript.m**. Now run (execute) the script. There are several ways to do this.

- Method 1: Type **MyFirstScript** into the Command Window and press ENTER.
- Method 2: Open **MyFirstScript** in the Editor and select the **EDITOR** tab from the toolstrip. Click on the green arrow labelled **Run**.
- Method 3: Click anywhere in the script window for **MyFirstScript**, and press **F5** on your keyboard.

Whichever method you choose you should now have an empty workspace and no figures open.

Next, we will add a new section to the script. The main advantage of using sections is that each section can be executed separately which helps with both organizing your code and with debugging. To add a new section, type %%. You can add comments after this but must leave a space between %% and your comments.

```
%% Parameter values and equation
Vmax = 6; % Maximum rate of reaction
Km = 40; % Substrate concentration at which the reaction rate is Vmax/2
S = 0:100; % Range of substrate concentrations
v = Vmax.*S./(Km+S); % reaction rates
```

You can run just this section of the script by clicking anywhere in the section so that it is highlighted and pressing CTRL-ENTER on your keyboard (or CMD-ENTER is you are using a mac).

The last step is to add some commands to plot the function:

```
%% Plot function
plot(S,v) % plot of s against v
xlabel('Substrate (S)')
ylabel('Reaction rate (v)')
```

Your completed script should now contain the following code:

```
%    MyFirstScript
%    Plots a Michaelis-Menten curve
%    My name
%    Created: today
clear
close all
%% Parameter values and equation
Vmax = 6; % Maximum rate of reaction
Km = 40; % Substrate concentration at which the reaction rate is Vmax/2
S = 0:100; % Range of substrate concentrations
v = Vmax.*S./(Km+S); % reaction rates
%% Plot function
plot(S,v) %plot of s against v
xlabel('Substrate (S)')
ylabel('Reaction rate (v)')
```

Running the code creates four new variables (`Vmax, Km, S` and `v`) in the workspace and generates a plot. If you can't see the plot, try checking to see if it is hiding behind the MATLAB interface (on Microsoft Window use ALT-TAB to do this) or tabbed behind another window in the interface. You can dock the plot to the interface by clicking on the arrow in the top right-hand corner of the Figure Window ↘ .

4.4.3 Colour Coding in the Editor

The Editor uses colour coding in two different ways:

- Syntax highlighting: this enables you to quickly identify sections of comments, text strings and keywords such as `for` and `end`. It also provides a quick way of seeing if you have made a typing error such as missing the end quotes from a text string. You can read more about syntax colours by typing the following in the Command Window prompt:

  ```
  doc Check syntax as you type
  ```

- Highlighting potential coding errors: Any code you write in the Editor is automatically analysed using the Code Analyzer. The offending piece of code will either be underlined with a red or an orange line. An orange or red line will also be shown in the right-hand column (message bar) of the Editor.

If you want to change the default colour codes, you can reset them in the Preferences tool which is found in the **HOME** tab on the toolstrip. For more details on how to do this, type

```
doc('change color settings')
```

into the Command Window to find the MATLAB documentation on this topic.

4.5 Debugging Code: Code Analyzer

MATLAB has a set of tools to help diagnose errors when you are script writing. The first useful feature is the Code Analyzer in the Editor. It provides a message indicator box at the top right-hand side of the Editor that will contain warning labels and will turn red, orange or green depending on whether errors have been found. If any syntax errors (those that stop the code from running) are found, the box will turn red. If there are possible improvements to code syntax but no errors, it will turn orange, and if there are no errors and no possible improvements to the code, the box will be green. Hovering over the box produces a message with this type of information.

Any code you type into the MATLAB Editor is automatically analysed for both fatal errors and efficiency. Fatal errors must be corrected before the code will run; however, efficiency errors are suggestions on how to improve your code but will not stop the code from running. To explore the Code Analyzer further, copy the following code, which contains deliberate errors, into a new script file, and save it as `AnyErrors.m`:

```
% AnyErrors.m
scaled discrete logistic model
clear all
GrowthRate = 2.8
K = 1
Steps = 100;
N(1) = 0.1;
for loop = 2:Steps
    Ni = N(loop-1)
    N(loop) = GrowthRate.*Ni.*(1-Ni)
% end
plot(1:Steps,N, 'x-']
title('Discrete Time Logistic
```


4.5.1 Fatal Errors (Red Line)

The Code Analyzer should highlight three fatal errors in the **AnyErrors.m** code by underlying the errors with a red squiggly line and placing a red line in the scroll bar on the right of the Editor. Hovering the mouse over either the red line in the scroll bar or the underlined bit of code will reveal an error message.

- First correct the fatal error at the end of the plot statement on line 12 of the code: hovering over the red line tells us that we have used an invalid syntax and correctly suggests that we might want to replace the square bracket with a round bracket which you should now do.
- The next error highlighted suggests that you need to add an **end** statement to match the **for**. To fix this, remove the % sign in front of the **end** statement on line 11 of the code. The **end** statement should change from green to blue. We have included a **for** loop to illustrate the importance of **end** statements – loops are covered in more detail in ▶ Chap. 11.
- The final fatal error tells us that we have failed to terminate a character vector on line 13. We need to add a single quote mark. Another fatal error will appear if you don't also add in the closing bracket.

Notice that MATLAB has not picked up the very first error in the code: the missing comment sign, %, on the second line. This is because until you try to run the code, it does not know that there is no function called scaled. Try running the code – an error message will appear in the Command Window (note you will only get this detailed message if you have saved the script).

```
Undefined function or variable 'scaled'.
Error in AnyErrors (line 2)
 scaled discrete logistic model
```

Any underlined sections of this error message can be selected with the mouse. Clicking on **AnyErrors** in this message opens the script file if it is not already open in the Editor. Clicking on **line 2** takes you to the line (in this case line 2) of the code in which the error has been identified. This is useful if you have a very large piece of code or a piece of code that calls other scripts and functions.

Correct the code by adding a % at the start of the second line.

4.5.2 Efficiency 'Errors' (Orange Line)

Once you have corrected the fatal errors and there are no red warnings, you should be able to run the code. Returning to the script you will still see some underlying orange lines and orange lines in the right-hand scroll bar. Four of these are because we have failed to terminate statements with a semi-colon. The semi-colon suppresses the printing of the output for that line of code to the Command Window. Allowing everything to be printed to the Command Window will increase the run time.

Now hover with your mouse over an orange line: a pop-up will appear with an explanation of why the Code Analyzer thinks you should include a semi-colon, and a '**`Fix`**' button. Click on the **`fix`** button and a semi-colon will automatically be inserted in the correct place.

You should still have one orange warning next to lines 3 and 10 of the code. Hovering over the warning for line 3 tells you that **`clear all`**

» usually reduces code performance and is often unnecessary.

Clicking on the **`details`** button provides a list of alternatives – you can either choose to ignore the warning or follow one of the suggestions. For now, replace **`clear all`** with **`clearvars`**.

Hovering over the warning on line 10 tells you that you are missing a semi-colon, which you can get the Editor to fix automatically, and that the variable **`N`** is changing size on every step of the loop. In most programming languages, you must declare the size of any array before assigning values to it. MATLAB allows you to build up an array as you go along, which can be useful if you don't know how big an array is going to be; however, it is computationally inefficient.

The impact of not pre-declaring **`N`** can be explored using the **`Run and Time`** tool which you can find on **`EDITOR`** toolbar in the toolstrip. Click anywhere in the script **`AnyErrors.m`** to make sure it is selected, and then click on the **`Run and Time`** icon. This will run the code and produce a pop-up window (which may be hidden behind the interface) with the total run time of the program and a breakdown of how each function call contributed to the total time. If you repeat **`Run and Time`** a few times, you will probably find that the total time varies a little according to what else is running in your computer background. Write down the total run time.

Now edit the code, adding in an extra line which declares the variable **`N`** as a (**`Steps`** by 1) vector of zeros. The complete code should be:

```
% AnyErrors.m
% scaled discrete logistic model
clearvars
GrowthRate = 2.8;
K = 1;
Steps = 100;
N = zeros(Steps,1);
N(1) = 0.1;
for loop = 2:Steps
    Ni = N(loop-1);
    N(loop) = GrowthRate.*Ni.*(1-Ni);
end
plot(1:Steps,N, 'x-')
title('Discrete Time Logistic')
```


Save the modified code and then execute it by using `Run and Time`. The code should have a shorter total run time.

4.5.2.1 Suggested MATLAB® Documentation Files to Investigate

To find out more about improving the run time of your code, search for the documentation:

```
doc preallocation
```

4.5.3 Sending Meaningful Output to the Command Window

Sometimes when you run a script, it is helpful to output some of the generated values to the Command Window. One way to do this is to combine the built-in functions `disp` and `num2str` (which converts a number to a string). To illustrate this, add the following line to the end of the `AnyErrors.m` script, and run the script again:

```
disp(['Population size after ',num2str(Steps),...
' steps =  ',num2str(N(Steps))])
```

4.5.3.1 Suggested MATLAB® Documentation Files to Investigate

To find out more about how to write meaningful output, search for the documentation:

```
doc disp
doc num2str
```

4.5.4 Executing a Subsection of a Script File

Rather than rerunning the whole script, sometimes you will want to run a few lines at a time. In the Editor you can do this by highlighting the lines that you want to run and pressing **F9** (or 'up arrow F7' on a Mac). Alternatively, right-clicking on the highlighted commands will reveal a pull-down menu from which you can select the option **Evaluate Selection**.

4.6 Saving Output from a Script File

All variables created by a script file, and not cleared within the script, are stored in the current workspace until they are overwritten or cleared from the workspace by using a command such as **clear**. When working with large data sets or simulation models, it can take several hours or even days to process the data or generate enough replicates of your simulation. It is therefore useful to know how to save this output for later use. One option is to save the entire workspace by using the pull-down menu in the top right of the Workspace browser; however, it is generally more efficient to include a command in your script which saves all variables in the workspace, or specified variables at appropriate points in your code. This can be achieved by using the command **save** followed by the name you wish to assign to the saved output file. Unless otherwise specified the variables will be saved in a MATLAB® **.mat** file.

For example, suppose you want to save the time course data and the value of the growth rate for **AnyErrors.m** to a **.mat** file, so you can access this information later. You can do this by adding the following line to the script:

```
save(['AnyErrors.mat'],'N','GrowthRate')
```

Running the script again will now produce a **MyFirstScript.mat** file in the Current Folder browser (if you can't see the file, then type **pwd** into the Command Window to check the current working directory as this is where the file will be stored). The file can now be opened in two ways in MATLAB. The first option is to type

```
clear % this empties the workspace so can see impact of load
load('AnyErrors.mat')
```

in the Command Window. Alternatively, you can double-click the MAT-file in the Current Folder browser to load the **.mat** file.

It is also possible to store data in different files, for example, if we are running repeated simulations within a script. To do this we use concatenation to create different file names. To try this out, create a new script called **LuckyDip.m**, add the following code, save and run it:

```
% Luckydip
% generates a random number and save to file
for i = 1:3
    B = rand;
    save(['LuckyDip',num2str(i),'.mat'],'B')
end
```

Note that this code uses for loops which will be covered in ▶ Chap. 11. The code will generate three **mat** files labelled **LuckyDip1.mat**, **LuckyDip2.mat** and **LuckyDip3.mat**.

4.6.1 Suggested MATLAB® Documentation Files to Investigate

You can also save variables to other file formats. To find out more, search for the documentation:

```
doc save
doc saveas
```

4.7 Using the Debugging Environment

Sometimes you may write a piece of code that does not produce any warning errors but does not produce the expected results. In these cases, the debugging tool may help. Here, we introduce how to use the Editor to debug a program.

To debug your code, you need to place breakpoints where you want MATLAB to pause execution of the file. There are three types of breakpoints: standard breakpoints, conditional breakpoints and error breakpoints. Here we use standard breakpoints.

Create a new standard script file, and add the following code saving the script as **TestBP.m**. Leave the script open in the Editor. The script contains a **for** loop to create a sequence of values (see ▶ Chap. 11 for more information on **for** loops).

```
% Fibonacci sequence
clear N j
N(1) = 0;
N(2) = 1;
for j = 3:10
    N(j) = N(j-1)+N(j-2);
end
```

Down the left-hand side of the Editor, there are two light grey columns. The first column contains the line number of the code. The second column is referred to as the breakpoint alley. All executable lines of the script are indicated by dashes in the breakpoint alley. Clicking on any dash in breakpoint alley will place a standard breakpoint on that line, shown as a red circle. You can also place a breakpoint by clicking anywhere in the line where you want to place the breakpoint, and in the toolstrip tab labelled **EDITOR,** click on the **Breakpoints** icon, and select **Set** from the pull-down menu.

If there is a syntax error in the script or if the final changes to the script are not saved, the breakpoint placed will be grey instead of red. In this case, you need to fix any syntax errors and save the file before you can continue.

Set a breakpoint on line 6 of the **TestBP.m** script. Next, run the script: the program should pause at line 6 and place a green arrow to show where the program has paused. At the same time, the Run button will change to a Pause button in the Run section of the **EDITOR** tab.

Look at the values of **j** and **N** in the Workspace browser. Now look in the Command Window: the command line prompt should have changed from >> to K>>, and you will now be able to type and run commands from the Command Window while the script is paused. Try this: change the values in the variable **N** by typing after the K>> prompt:

```
N = [3 5];
```

Press ENTER and look at the workspace again – the value of **N** has been updated. Any commands typed in at the K>> prompt do not permanently change the code, and if you run **TestBP.m** again, **N** will default to the values assigned in the script. Click on **Continue** (in the **EDITOR** toolbar on the toolstrip) to get the program to loop until it reaches the same point again. What has happened to the values of **j** and **N**?

You can also step through the code line by line by selecting **Step** from the **Debug** section of the **EDITOR** toolbar. When you have finished, press **Quit Debugging** – a red square button on the **Debug** section of the **EDITOR** toolbar.

To clear a breakpoint, click on the red circle so that you get a black line again. If you want to disable a breakpoint without removing it, so that the program ignores it when running, right-click on the breakpoint, and select **Disable breakpoint** from the pop-up menu. A cross will appear over the red breakpoint. If you want to enable a disabled breakpoint, right-click again, and select **Enable** from the pop-up menu.

While the program is paused, you can also modify other sections of the code. However, you will need to save the program before you can continue the debugging process. It is not currently possible to debug a section within a script by running that section alone – any breakpoint placed in the section will be ignored by MATLAB unless you run the whole script.

4.7.1 Suggested MATLAB® Documentation Files to Investigate

To find out more, search for the documentation:

```
doc debug a matlab program
```

4.8 More Editor Tips

4.8.1 Code Folding

Code folding enables you to hide parts of the code to make it more readable, but it does not change how the script behaves. Open **TestBP.m** in the Editor. You will see a small box with a dash inside next to line 5. Extending from the bottom of the box is a line which goes to line 7, the end of the **for** loop. Clicking on the box will compress the code, so you can no longer see the contents of the loop. To explore code folding, further select the **VIEW** tab from the toolstrip. Here you can quickly fold and unfold sections of code or the entire script. From this tab you can also select features which allow you to customise the Editor such as tiling so that you can see more than one script file at a time or splitting the screen so that you can freeze a section of code while you scroll through the rest of the code.

4.8.2 Large Section of Comments

If you want to create several lines of comments, then it can get cumbersome to start every line with a %. Instead you can encase them using %{ and %}. In this case it is important that the opening and closing codes are on separate lines as in the example below. You can now 'fold' the large comment section.

```
% Fibonacci sequence
 clear N j
 %{
The Fibonacci sequence is frequently observed in biology
See for example
Novel Fibonacci and non-Fibonacci structure in the sunflower:
results of a citizen science experiment
Jonathan Swinton, Erinma Ochu, The MSI Turing's Sunflower Consortium
R. Soc. open sci. 2016 3
 %}
 N(1) = 0;
 N(2) = 1;
 for j = 3:10
     N(j) = N(j-1) + N(j-2);
 end
```

4.8.3 Adding Formatted Text to Your Code

Basic formatting, such as bulleted lists and LaTeX formatted equations, can be included in the comments section of a standard script file. These are useful if you want to produce a report from your script although they are somewhat superseded by live scripts, so we will not go through the options here. If you want to know more, search the MATLAB documentation for **`Publishing Markup`**.

4.8.4 Code that Is Too Long for One Line

If you need to write a long line of code, you can split it up by inserting ellipsis at the end of the first line before starting a new line – this tells MATLAB that the line isn't finished yet. For example, we might want to add a line to the **`TestBP.m`** script which prints out the Fibonacci sequence to the Command Window:

```
disp(['The first 10 numbers of the Fibonacci sequence are ',...
    num2str(N)])
```

4.8.5 Finding All Instances of a Variable

Clicking next to any variable name will highlight in light blue all instances where this variable occurs in the script. Try this for `N` or `j` in the `TestBP.m` script. This does not work if you click on the variables names in the `clear` statement line.

4.8.6 Housekeeping

Each time you open a new file in the Editor, you will notice that the other files remain: you can select any file by clicking on the relevant tab. When you have finished with a script, you can close it by clicking on the light grey cross next to the file name in the tab.

4.8.7 Keyboard Shortcuts

While most tasks can be achieved by selecting the correct icon in the menu, it can be useful to learn some of the more common keyboard shortcuts. Hovering over an icon will generally tell you if a shortcut is available. Shortcuts for running code are: F5, which runs the whole script; F9 ('up arrow F7' on a Mac), which runs any lines of the script that have been highlighted using the mouse; and CTRL-ENTER (or CMD-ENTER on a mac) which runs the highlighted section of a script.

A complete list of keyboard shortcuts can be found by navigating to the `HOME` tab on the toolstrip, finding the Environment section, and clicking on Preferences. Within the Preferences window, select `MATLAB > Keyboard > Shortcuts`.

4.9 Using the Live Editor

Introduced in release 2016a, the Live Editor enables you to create scripts which combine formatted text, graphics and equations with sections of code. This allows you to produce interactive documents that can be used for teaching purposes; producing reports; and, to help you document your work. Output from live scripts is displayed within the script in the Editor next to the code which generated it. Live scripts can be converted to HTML, PDF or LaTeX files for publication.

To create a live script, you can use any of methods described for the creation of a standard script, but this time select Live Script for the first two methods (see the earlier section in this chapter: Creating and Saving a new script file). Live script files have the extension `.mlx`.

You can also convert a standard script to a live version. Use one of the following methods to open the `AnyErrors.m` file as a live script:

- **Method 1**

Navigate to the `AnyErrors.m` file (which should be in in your `MyTestScripts` folder) in the `Current Folder` window. Right-click on the file name: this brings up a menu from which you should choose `Open as Live Script`. Your file will now open in the Editor as a live script. Save the file as `Logistic.mlx`

Method 2

If you already have the `AnyErrors.m` file, open in the Editor, and then you can click on the tab for that file name in the Editor window so that it is moved to the top of the Editor files and its contents are visible. Now right-click on the file name in the Editor: this will reveal a menu from which you should select `Open AnyErrors.m as a Live script`. Save the file as `Logistic.mlx`

Warning!
Changing the extension from .m to .mxl will not convert a standard script to a live script.

Look at the live script `Logistic.mlx` that you have just created in the Editor. You should see that all the code from `AnyErrors.m` is now encased in a lightly shaded box and the window is split vertically with nothing to the right. In the top of the scroll bar on the right of the Editor, there are two small icons which allow you to switch between displaying output directly below sections of code written or on the right-hand side of the code. You should also notice two new tabs have appeared in toolstrip menu labelled `LIVE EDITOR` and `INSERT`. Click on the `LIVE EDITOR` tab so that you can see the commands for the Live Editor.

Notice the sections labelled `TEXT` and `CODE`. If you are currently in code mode, then you can switch to text mode by clicking on the text icon. The keyboard shortcut for this can be found by hovering over the `Text` icon (ALT + ENTER on Microsoft Windows).

Now add some text to `Logistic.mlx`. One way to do this is to use the mouse to select the commented section at the start of the code and then click on the text icon. You can format the text using the options in the `TEXT` tab. If you want to include any images or equations, select the `INSERT` tab where you will find these options. Equations can be typed in using either the equation interface or via a pop up window which can interpret LaTeX commands.

Now run your code by clicking on the `Run` icon in the `LIVE EDITOR` tab. The method you use to enter debugging mode in Live scripts depends on which version of MATLAB you have. In MATLAB r2019a you can use the 'run to here' tool which appears if you hover to the left of any line of code. In MATLAB r2018b you can set breakpoints by clicking on any line number. This will put a green arrow on the line number. If you right-click on this arrow, a menu will appear with breakpoint options. Running the code now will send it into debugging mode.

Remove any breakpoints and run the code. The code will generate a graph next to the section of code with the `plot` command. You can open the graph in a Figure window for editing by clicking on the arrow in the top right-hand corner of the figure that appears when you hover the mouse over the image as shown in Fig. 4.2.

Another advantage of the Live Editor is that you can make your code interactive by including drop-down menus and control bars which allow the user to select parameter values. Suppose we want to explore the impact of the growth rate on population dynamics. To do this use the mouse to select the numerical value next to the `GrowthRate` (currently set as 2.8). Now click on the `INSERT` tab on the toolstrip, and choose `Numeric Slider` from the drop-down menu below the `Control` icon. This produces a box in the Editor in which you can insert a minimum value, step size and maximum value that you want to be able to select. Set these as 0.1, 0.1 and 5, respectively. If you can't see this slider properties box then right-click on the slider and select to Configure Control from the pull down menu.

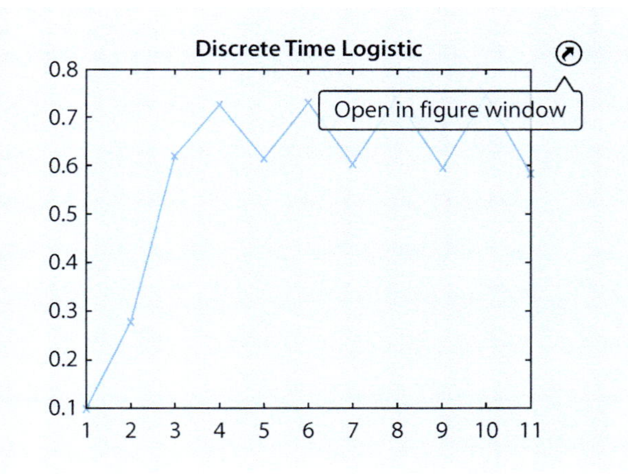

Fig. 4.2 Example of a figure window within the Logistic.mlx file where the mouse has been hovered over the arrow in the top right-hand corner

You can now select the value of the growth rate by moving the slider. Each time you move the slider, that section of the code will rerun, and, if your graph is in the same section of code, then the figure will also be updated.

Once you are familiar with the Editor, you can easily switch to working in the Live Editor. All the functions and the hints we have given can be translated to a live version. One current disadvantage of working in the Live Editor is that it may increase the total run time.

Take-Home Messages

Script files provide a way of saving a group of related commands in one document. An important part of script writing is documentation: always try to make sure you add comments to your code so that when you return to the code or share it, it is clear what your script is doing and why you have used certain commands.

Dividing script files into sections allows you to further organize your work and helps to make your script readable and easy to check. Now that you are familiar with how to use the Editor to create script files, we recommend that as you work through the rest of the book, you get used to writing your code in script files (either live or standard scripts) and avoid using the Command Window except for quick calculations or to check how individual functions work.

Accessing Help

C. R. Webb, M. Domijan, *Introduction to MATLAB® for Biologists*,
Learning Materials in Biosciences, https://doi.org/10.1007/978-3-030-21337-4_5

What You Will Learn in This Chapter

As you begin to develop and adapt code for your own projects, it is inevitable that you will come across questions which we have not answered in this book. The options for refining MATLAB® code and graphics are so vast that MATLAB no longer produce printed copies of manuals. Instead the manuals have been replaced with comprehensive documentation files and online support. In this chapter you will learn how to find help both within the MATLAB interface and online at the MathWorks® website. You will learn how to access free-to-use code shared via File Exchange on the MathWorks website and how to use Cody™, a MathWorks game environment in which you can test and develop your programming skills.

5.1 MathWorks® Account

Almost everything you need to know about MATLAB can be found either within the program documentation or via the MathWorks website. MathWorks, Inc. is the company behind MATLAB. Before you continue with this chapter, we recommend that you create an account with MathWorks if you don't already have one. You will not be able to download files from the MathWorks user community without an account. To set up an account, visit the MathWorks website, ► https://www.mathworks.com, and select `sign in` – you will be prompted to create a new account if you don't have one.

There is no cost involved, and you do not need to have a MATLAB licence to set up an account. Once you have set up an account, you retain the same account for life. You can link your MATLAB licence to your MathWorks account by providing a valid activation key or licence number.

5.2 Backward Compatibility

It is beneficial to know which version of MATLAB you are using: if possible you should install the latest version. The MATLAB package is updated twice a year, with new functions and features added with each release. In general, MATLAB is backward compatible; in other words, you should be able to use code written in an older version of MATLAB® than your version without any issues. You may however encounter run time errors if your version of MATLAB® is older than the source code and the source code contains functions released after your version of MATLAB. You can only use Live Scripts from 2016a onwards. To find out which version of MATLAB you have, and which toolboxes are included in your licence, type:

```
ver
```

A record of changes made in each update of MATLAB is given in each version's Release Notes. These can be found on the MathWorks website and in the documentation browser:

```
doc MATLAB release notes
```

5.3 Overview of Help Options

There are a wide variety of help options for MATLAB users. These can be broadly categorized as follows:
- Offline help direct from the MATLAB program (available if you installed the documentation when you installed the program on your computer)
- Online help via the MathWorks website
- Online help from forums not directly linked to the MathWorks website

5.3.1 Offline Help Within MATLAB®

For the purposes of this section, we will assume that you have downloaded the help documentation when you installed MATLAB on your computer. If you did not do this, do not worry – all the help documentation is available online on the MathWorks website, so if you are connected to the Internet, you should be able to do most of the tasks described in this section.

The MATLAB interface provides several ways to access help from two distinct sets of documentation files: interactive **doc** files and legacy **help** files. How you access help will depend partly on the type of help you need. We have broadly classified them as:
- Exploring the documentation files
- Obtaining help for a specific function
- Finding a function to perform a specific task
- Browsing for an appropriate function

5.3.1.1 Exploring the Documentation

The MATLAB developers have tried to make it as easy as possible to find the help you need. As a result, there are several ways you can access the documentation pages: you may even find some that we haven't mentioned here. The first option is to call them from the Command Window using the keyword **doc**:

```
doc
```

This opens a window labelled **Help** which contains a tab opened on **MATLAB Documentation**: this page is like the contents section of a book, and from here you can search for specific terms or explore the many ways that MATLAB can be used. You can also try out some of the examples and tutorials which illustrate various aspects of the main program. You may see a list of toolboxes, if you installed any with your copy of MATLAB, and links to the documentation for functions included in these toolboxes. You can dock the **Help** window to the main user interface by clicking on the arrow ↘ in the top right corner of the window. If you have not downloaded the complete documentation, you may get a **page not found** error. In this case you can open the online documentation by selecting the **HOME** tab on the toolstrip and then selecting **Support Web Site** from drop-down list below the **Help** icon in the **RESOURCES** section.

Another way to access the same documentation page is to click on the question mark icon which can be found both near the top right-hand side of the user interface and in the **HOME** tab of the toolstrip in the **RESOURCES** section. You can also use the keyboard shortcut, function key F1, to access the documentation from the MATLAB interface.

5.3.1.2 Obtaining Help for a Specific Function

There are several ways you can access help and documentation for a specific function. Suppose, for example, that we want to find the sum of some values in a matrix that we have previously defined and we already know that the function we want to use is called **sum**.

- **Option 1: Start Typing the Command in the Command Window or the Editor**

To try this out type the following into the Command Window or a standard script file, pausing after you have typed the final character:

```
MyData = randi(100,6,4)
sum(
```

Pausing a short while will bring up a pop-up window with a concise list of the required and optional input arguments. If you need to find out more about how to use the arguments, you can click on the **More Help**... link. This will open a window with the documentation file for the function.

Sometimes the pop-up window does not appear, or disappears too quickly, in which case place the cursor next to the bracket and press the function key **F1** on your keyboard. This will take you straight to the documentation for that function.

In Live Scripts the initial pop-up has been modified to provide a more concise overview of the function. The information displayed changes as you add inputs to the function. To explore this, open a new live script, and making sure you are in code mode (as opposed to text mode – see ▶ Chap. 4 if you are unsure how to do this), type in:

```
MyData = randi(100,6,4)
sum()
```

Now type the letters **My** between the brackets next to the **sum** command as follows:

```
sum(My)
```

A pop-up window will appear with a list of suggestions to autocomplete this text string[1]. Select **MyData**. Next type a comma so you have

1 Note: this is a relatively new feature of Live Scripts and may not work in earlier versions of MATLAB® (e.g. 2016b).

```
MyData = randi(100,6,4)
sum(MyData,)
```

Another pop-up will appear this time suggesting inputs for the dimensions to operate on (i.e. do you want the sum of each row, each column or the overall sum). Select `'all'` and type in another comma so you now have

```
MyData = randi(100,6,4)
sum(MyData,'all',)
```

Pausing again will give a pop-up window with additional options such as the option to include **NaN**[2] in the calculation. None of these options are useful here so delete the second comma and execute the code. At each stage the pop-up windows also include the question mark icon which you click on to get to the main documentation page for the **`sum`** function.

▪ Option 2: Open Function-Specific Documentation (doc) Files

There are two main routes to finding the documentation files for a specific function from the MATLAB interface. The first is to use the command line:

```
doc sum
```

If you are looking for documentation on something which has more than one word, you may need to include the search term in single quotes:

```
doc('Random Number Generation')
```

The second option is to use the box in the top right-hand corner of the main interface, which by default contains the text **`Search Documentation`** in light grey. When you type **`sum`** into this box, a pull-down menu will appear with the functions which have the closest match to the typed function name. Clicking on any of these will take you to the relevant documentation page.

▪ Option 3: Open Function-Specific help Files

Concise help files can be accessed from, and displayed in, the Command Window. These help files provide a quick overview of the specified function and, in contrast to the doc files, are also accessible within MATLAB® Mobile™. If you use help from the desktop application, you can prevent the file from scrolling too quickly by first using the command **`more on`**.

2 NaN is short for Not-a-Number and can be used to represent missing data.

```
more on % turn off using more off
help sum
```

Clicking on any of the **see also** links (e.g. **cumsum**) at the bottom of the help file for **sum** will take you to the help file for that function. The end of the help page gives two further clickable links. The first link, **Reference page for sum**, takes you to the more detailed documentation which we have already found by typing **doc sum**. The last link labelled **Other functions named sum** provides links to other help files which have arisen because you can use the **sum** function for different data types and help pages have been constructed for each of these data types.

5.3.1.3 Finding a Function to Perform a Specific Task

Suppose you want to add some error bars to a plot and want to know if there is a function to facilitate this. You can search the help files from the Command Window using the built-in function **lookfor**:

```
lookfor('error bars')
```

This will bring up a link to a function called **errorbar**. This method can be a little slow, and it is usually quicker to search the browser-based documentation using the command

```
docsearch('error bars')
```

This takes you to the help documentation window and lists possible pages that match your search term. You can filter the results so that it only shows functions by clicking on **FILTER** in the top left of the results page. You can achieve the same result by typing **error bars** into the box in the top right-hand corner of the main interface which has the words **Search Documentation** in light grey and either choosing the most appropriate function from the pop-up menu or pressing ENTER to go to the same page as we found with **docsearch**.

You can also find functions in Live Scripts using the autocomplete option[3]. To explore this open a Live Script and in code mode type **err**. A list of functions starting with the letters **err** will appear from which the function you want may be obvious (in this case **errorbar**). Clicking on the appropriate function will add it to your script. Now you can access help on the function by: pressing F1 on your keyboard; right-clicking on the function name and selecting **Help on "errorbar"** from the pull-down menu; or by typing an open bracket after the function name and waiting for the pop-up help to appear as described in a previous section.

3 This feature is only available in recent versions of MATLAB.

5.3.1.4 Browsing for an Appropriate Function

If you vaguely know what you want to do but aren't quite sure what is available, you can either explore the main documentation as described at the start of this chapter, or you can use the Function browser. The Function browser provides a clean view of the functions available under various headings. To access this, click on the *fx* icon which is in the left-hand column of the Command Window by the command prompt. This brings up a searchable list of functions available.

Using the Function browser, you can locate the `sum` function by selecting the following hierarchy of links: MATLAB, Mathematics, Elementary Math and Arithmetic. This will give a list of related functions including the `sum` function. Click on `sum` to reveal a pop-up window which contains the basic syntax and a clickable link labelled `More help...`.

5.3.2 Navigating Around the Function Documentation Page

The precise structure of the documentation file varies between functions. For instance, suppose that we want to understand the function `sum` in more detail. First open the documentation for `sum` (e.g. type `doc sum` into the `Command Window`), and dock the `Help window`.

The documentation starts with a brief definition of the function. This is followed by a series of sections which you can navigate to by following the links on the left-hand side of the document page:

- `Syntax` section summarises the different ways in which the function can be used. In the case of the `sum`, there are six different ways.
- `Description` section gives a more detailed description for each syntax option and a hyperlink to an example for each option. Select the link next to the last syntax option. This takes you to an example entitled `Sum Excluding NaN`. Including a `NaN` value means that your sum will be `NaN` if your matrix contains any missing values. To explore this in more detail, you can select `"Open Live Script"`. This will open a live script in the Live Editor from which you can run the example code. You can edit this file to explore the impact of including `'omitnan'` as an option. Be careful not to save this file without changing its name as this will overwrite the example file permanently in your copy of MATLAB.
- `Examples` section contains all the examples which we could link to from the `Description` section.
- `Input Arguments` section, when each heading is expanded, gives detailed information on the type of input possible and how the function uses them. Take some time to read the section on dimensions as it applies to a lot of functions that will be useful to you.
- `Extended Capabilities` section provides additional information on the use of the function. If you are dealing with very large data, you may be interested in exploring `Tall Arrays`.
- `See Also` section can be particularly useful in finding other related functions which may be more applicable to your problem.

Occasionally you will come across functions which have been superseded but not removed to retain backward compatibility. For example, type in the Command Window

```
doc ezplot
```

This brings up the documentation on a plotting function which was particularly useful for plotting circles. Before the syntax is introduced, a warning is issued that the function **`is not recommended`**, and another function, **`fplot`**, is suggested as a replacement. Both methods can be still be used to plot a circle:

```
ezplot('x^2+y^2=1')
figure
fplot(@(t)(cos(t)), @(t)(sin(t)),[0,2*pi])
```


5.3.3 Online Help: MathWorks® Website

As well as providing access to the MATLAB documentation, the MathWorks website provides a wide range of resources to help both novice and experienced users. It is worth dedicating some time to exploring the website so that you are aware of the range of help and tutorials available. Here we will focus on two areas of the website: **`File Exchange`** and **`Cody`**™ which are freely available to all MATLAB users. To find these sections of the MathWorks website, you can navigate directly from MATLAB by selecting the **`Community`** link which can be found in the **`RESOURCES`** section of the **`HOME`** tab on the toolstrip and then choosing the relevant section from the website.

5.3.3.1 File Exchange ▸ mathworks.com/matlabcentral/fileexchange

File Exchange is a repository in which users can share code, apps and examples for anyone to use without charge. The shared code ranges from short scripts and files to perform relatively basic tasks to packages of functions which form specialist toolboxes. There are many useful discipline-specific examples and toolboxes which could save you from having to write complex code.

Suppose, for example, that we want to see if anyone has submitted an alternative to the MATLAB built-in **`errorbar`** function. Enter **`ErrorBar`** into the **`Search Files`** box on the File Exchange webpage. This will give a list of File Exchange contributions which contain the search term. You can choose to sort contributions by various factors such as contribution date, user rating or number of downloads.

Each contribution has a user rating (up to 5 stars), and you can see how many users have assigned a rating to the file, how many downloads there have been in the last 30 days and how many comments have been made. Select one of these files by clicking on the name of the file (e.g. **`raacampbell/shadedErrorBar`**). This will take you to a page which provides a more detailed overview of the function. Here you can also read any user comments together with responses from the authors. These can be particularly useful if you are having problems understanding how to use the functions as someone may already have asked the same question. You can also click +Follow which allows you to be informed via e-mail of any changes that happen to this file (such as updates that the contributor might make).

If you want to investigate the function further, you can now press the download button. You will only be able to download the function if you are logged into MathWorks. Unzip and copy the files to a suitable directory, and make sure they are visible in MATLAB (by using **`addpath`** or right-clicking on the folder in the **`Current Folder`** window and selecting **add to path> selected folders and subfolders**). Some contributions include a demo file so that you can immediately try out the function. Many also include a **readme** file with more information on how to use the function and a **licence** file. Typing **doc** into the **Command Window** followed by the function name will open a help file on the function which may contain helpful information on how to use the function.

Be aware that as with any freeware, these files don't necessarily go through the rigorous testing that built-in functions do. You can look at the number of the people who rate the file and the comments of the other users, but ultimately it is your responsibility to try to check the code to make sure it does what you think it should do. You should also acknowledge the authors of any code you use in your publications.

If you want to explore some of the more interesting contributions to File Exchange, a good starting point is to scroll through the File Exchange Pick of the Week blog ► https://blogs.mathworks.com/pick/ which reviews some serious and some less serious contributions to File Exchange.

5.3.3.2 Cody™

Cody™ is a MATLAB problem-solving game that allows you to develop and refine your MATLAB coding skills. The challenges are a mixture of those developed by the Cody MATLAB team and user-contributed challenges. Challenges vary in difficulty with many suitable for novice users of MATLAB. To illustrate how to use Cody, we will explore one of the simpler problems. First navigate to the Cody section of the MathWorks website – you need to be signed into MathWorks to be able to use Cody. From here select **Groups** and find the problem group called **Cody Challenge**. Scroll down until you find the problem labelled **"Times 2 – Start here"**. This is the first problem in the Cody Challenge. Click on the problem name to get to a page explaining what the challenge is.

» Given the variable x as your input, multiply it by two and put the result in y.

To attempt this challenge, click on the **solve** button to start your solution. This will bring up a section of code for you to edit.

```
function y = times2(x)
  % Modify the line below so that the output y is twice the
% incoming value x
  y = x;
  % After you fix the code, press the "Submit" button, and you're
% on your way.
end
```

The output value is **y**, and the input value is **x**, so if we want to multiply the input by 2, we need to modify the third line of code. We do not need to alter the first and last line of code. Amend the code inside the function as follows:

```
y = 2.*x;
```

Now press the submit button. You should see a green tick to say you have correctly solved the problem. You will also see a statement of how big your solution is (solution size) and how small the best solution submitted is. You can compare the efficiency of your solution with others submitted by selecting `View on Solution Map`. You can click on the solution map to see how other users solved the problem. This can be a useful way of seeing alternative, perhaps more elegant, ways of getting to the same solution.

5.3.4 Online Help: Other Sources

Often the quickest way to answer a MATLAB coding question is to type your question into a standard search engine, always remembering to use the keyword MATLAB in your search string. Frequently someone else has already asked the same question, and there is at least one solution available. In general, the first few links will take you to various pages on the MathWorks website. You may also find links to forums not hosted by MathWorks which can provide informative advice.

5.4 Potential Issues with Shared Code

Updates of MATLAB are issued on a 6-monthly cycle. They are called 'a' and 'b'. If you are using an out-of-date version, then some functions written in a newer version may not exist in your version. If you are using a very old script, you may also encounter errors as some functions have become obsolete. Sometimes MATLAB will also warn you if want to use a built-in function that is about to be replaced by another. Another reason why someone else's code may not work for you is if you are using a script written by another person who has access to different toolboxes and hence different functions to you (see ▶ Chap. 6).

Take-Home Messages

There are many ways in which you can find advice and help both within the MATLAB® interface and online at the MathWorks® website. Spending some time now familiarizing yourself with the support available will save you many long nights of trying to solve coding problems for which there is an existing solution. For many problems, a web search of your question will take you to either a community-supplied solution or a MathWorks solution or occasionally will confirm that whatever you are trying to do is impossible. We also recommend returning to Cody™ as you progress through this book to help develop your understanding of MATLAB.

Exploring Built-In Functions

C. R. Webb, M. Domijan, *Introduction to MATLAB® for Biologists*,
Learning Materials in Biosciences, https://doi.org/10.1007/978-3-030-21337-4_6

What You Will Learn in This Chapter

MATLAB® is a structured programming language – this means that you do not have to code everything from scratch. This makes it easier and quicker to develop sophisticated code. The program contains a large library of built-in functions, many of which are included in the core package available to all users. In this chapter, you will learn how to interpret function syntax, as provided in the help documentation, so that you can make optimal use of these functions in your own work. By the end of this chapter, you will have explored a variety of function structures including functions with optional inputs and multiple output values. You will also learn how to use a range of functions including the **sub2ind** and **ind2sub** function to convert between the subscript and index method of identifying location in an array, the **find** function to select a subset of elements in an array, and the **max** function which is a flexible function used to find maximum values.

In the last section of the chapter, we introduce methods to help you identify whether a function belongs to the main program or a toolbox: toolboxes consist of sets of functions which focus on a specialist area, and you can only use these functions if you have a licence for the relevant toolbox.

6.1 What Is a Built-In Function?

A function is a set of commands stored in a program file that can be called by any other script or from the command line. If you are following this book in chapter order, then you will have tried out several commands such as `ones`, `zeros` and `eye`. These are all functions which come with a basic installation of MATLAB®. We define **built-in functions** as functions that are supplied as part of the MATLAB core installation or with any of the MATLAB toolboxes. We refer to MATLAB as a structured programming language because it has a large library of built-in functions that make it easier to write code than in a basic programming language.

Functions versus script files: Any script file or function in the current path can be called by any other script file or function. The first big difference between scripts and functions is that functions can have input and output parameters. **Script** files can only operate on the variables that are hard-coded into their m-file. The second important difference is that the variables created and used within a function do not automatically go in the main workspace unless this is specified in the code.

In this chapter we will explore how to apply the function syntax to enable you to make the most of built-in functions. Plotting functions are discussed separately in ▶ Chap. 7. We show you how to write your own functions in ▶ Chap. 9.

6.2 Exploring Function Inputs and Outputs

We have divided this section into the main types of functions you may encounter in ascending order of complexity, introducing functions which we find particularly useful along the way:

- Functions that do not require any inputs
- Functions with a fixed number of inputs

- Functions with optional inputs
- Functions with two or more outputs
- Functions with optional inputs and outputs
- Applying functions to data 'structures'

6.2.1 Functions that Do Not Require Any Inputs

Not all functions require inputs to produce an output. Try the following examples (either type them one at a time into the Command Window; or put them in a script, run it and look at the output in the workspace or Command Window; or put them in a live script and run it):

```
Func1 = NaN
Func2 = rand
Func3 = datetime
Func4 = eye
Func5 = zeros
```

The functions **NaN**, **rand**, **datetime**, **eye** and **zeros** are examples of functions which have a default output if no input is supplied. The output can be changed by supplying one or more inputs; try, for example:

```
Func1b = NaN(2,3)
Func2b = rand(10,3)
Func3b = datetime('tomorrow')
Func4b = eye(5)
Func5b = zeros(2,4,2)
```

There are a small number of functions which cannot take any inputs. One example is the function for the mathematical constant π:

```
pi
```

Functions can be included within longer calculations. For example, suppose we wanted to find the area of a circle with specified radius – we can insert the function name **pi** into our equation:

```
r = 3; % radius
Area = pi.*r.^2
```

6.2.2 Functions with a Fixed Number of Inputs

When you write your own functions, it is likely, at least initially, that you will create functions that require a fixed number of inputs (see ▶ Chap. 9). There are many built-in functions which follow the same format. Try out the following examples:

```
A = [4 5 -8; -5 -6 9; -2 -2 3];
disp(A) % print out matrix A to the command window
det(A) % determinant of matrix A
transpose(A) % flip the matrix about the diagonal
inv(A) % inverse of matrix A
power(A,3) % cube each value in the matrix
diag(A) % extract the values in the main diagonal of the matrix
```


For detailed explanations of each of the above functions, open the MATLAB documentation for the function using the command **doc *function_name***, for example, **doc transpose** (see ▶ Chap. 5 for how to interpret the MATLAB documentation for a function). Many functions can be applied to arrays of data such that the function is applied separately to each element in the array. Using functions in this way can save the need for more complex programming using loops (see ▶ Chap. 11 to find out more about loops). Try for example:

```
SomeSquares = [1 4 9; 16 25 36];
sqrt(SomeSquares)
```

6.2.3 Functions with Optional Inputs

Many built-in functions have one or more optional inputs. This gives the user greater flexibility in how the function can be used and reduces the number of separate functions (and hence function names) that need to be developed for closely related tasks. Consider, for example, the function **randn**. This function generates a random number drawn from a normal distribution with mean 0 and standard deviation of 1. Suppose we want to generate a host landscape on a grid whereby each cell is assigned a randomly selected value drawn from a normal distribution with mean of 15 and standard deviation of 2. One option might be to set the value in each grid cell in turn: to do this we can use the **randn** function with no inputs and scale the output so that we have the correct mean and standard deviation:

```
mu = 2;
sd = 15;
oneHost= mu*randn+sd
```

We could now use a **for** loop to assign a value in each cell (see ► Chap. 11). Far simpler, however, would be to provide an input to the **randn** function to generate an array of random values, as in the following two examples:

```
HostLandA=2*randn(5)+15
HostLandB=2*randn(5,3)+15
```

HostLandA is a 5 by 5 matrix of cells with each cell containing a value drawn from the normal distribution with mean 15 and standard deviation 2. Providing a single input value to the **randn** function has generated a square array of data with the number of rows = number columns = input value. **HostLandB** is a 5 by 3 array of elements. As two inputs were provided, the first input is now interpreted by the function as the number of rows in the output array and the second input as the number of columns. Suppose we want to generate four different 5 by 3 landscapes. One way to do this is to add an extra dimension to the **randn** call:

```
HostLandC=2*randn(5,3,4)+15
```

To access the second layer of the landscape, type, for example:

```
Landscape2 = HostLandC(:,:,2)
```

You can keep adding comma-separated integers to the input list in **randn** to create a multidimensional array. An additional option is to specify whether the output is of numeric type double or single precision by adding an extra input (the default is double, and for most scenarios, there is no reason to change from this):

```
HostLandD = 2.*randn(5,3, 'single')+15
```

Many functions have one or more core inputs and one or more optional inputs. To explore how to distinguish between them, we will explore the built-in function **mean**. Type the following into a script or the Command Window, and pause for a short while (don't press ENTER):

```
mean(
```

A list of possible inputs and the option to open the help file will appear. The first input is colour-coded blue to show that it is a required input. We are then presented with a list of other input options and alternative ways in which the function can be used (the precise layout of these options depends on which version of MATLAB you are using). Apply the **mean** function to one of the host landscapes:

```
MuHLB = mean(HostLandB)
```

The output, **MuHLB,** is a row vector with three values. The entries of this row vector are the mean values for each columns of the input array. We would expect these to be close to 15 (check you understand why). Typing


```
ColMuHLB = mean(HostLandB,1)
```

gives the same answer. This is because the second input specifies the dimension along which the mean will be calculated and typing **1** as the second input specifies that the mean will be taken for each column. This is the default setting, which is used when we don't explicitly define the optional input. On the other hand, typing

```
RowMuHLB = mean(HostLandB,2)
```

specifies that the mean should be calculated along the rows, so the output is a 5-element column vector. If you want to find the overall mean, there are three options: (i) take the mean of the row means (which we calculated above and stored as **RowMuHLB**); (ii) you can avoid creating a separate variable for the row means and instead nest a **mean** function within another **mean** function; or (iii) you can use the colon operator to turn the matrix **HostLandB** into a single column vector and find the mean of the vector:

```
mean(RowMuHLB)
mean(mean(HostLandB))
mean(HostLandB(:)) % turns matrix into a column vector first
```

There are two other optional values in the **mean** function which can be used to specify the output type and to determine how the function should be applied if there are missing values (which must be included as **NaN**) in the input. This second option can be useful if you have experimental data with missing values:

```
MyHostData = [14 12 10 NaN 16];
mean(MyHostData)
mean(MyHostData,'omitnan')
```

The default for the **mean** function is to include missing numbers. If we include the command to omit missing values, 'omitnan', then the mean is calculated as the sum of all the valid data (i.e., not the **NaN** values) divided by the number of valid data points.

6.2.4 Functions with Two or More Required Output Values

Some functions are designed to produce two or more output parameters. One example is the **ind2sub** function. The **ind2sub** function is used to convert an index value for a given size of matrix into the subscript location for the same size matrix and so has two outputs: the row number and the column number. These two methods of locating a value in an array of data were introduced in ▶ Chap. 3 (see ◘ Fig. 3.2). The function **sub2ind** works in the opposite direction converting a subscript (i.e. row, column) position to an index position and produces a single output. To explore these functions further, start the example by copying the code below into a script to produce a 5 by 5 times table square:

```
%% create a times-table square
FiveTT = [1:5].*transpose([1:5]);
```

Run the code to create the variable **FiveTT** in your workspace. If we want to know what 3 times 5 is, the simplest way would be to use subscript notation to extract the value found by navigating along the third row to the value in the fifth column (or we could choose the fifth row and third column):

```
FiveTT(3,5)
```

We can convert the subscript location using the function **sub2ind** by supplying the size of the matrix, the row number and the column number:

```
Ind3T5 = sub2ind(size(FiveTT),3,5)
```

Now we can convert back to a row and column number by using **ind2sub**, but this time we need to supply two output names encased in square brackets and two inputs (the size of the matrix and the index value):

```
[rowLoc, colLoc] = ind2sub(size(FiveTT),Ind3T5)
```

You can check that these give the same value in the matrix:

```
FiveTT(Ind3T5)
FiveTT(rowLoc,colLoc)
```

Suppose now we create a larger times table square:

```
SevenTT = [1:7].*transpose([1:7]);
```


If we now compare the result we get using the subscript location and the index location we just found, we will get different answers:

```
SevenTT(rowLoc,colLoc)
SevenTT(Ind3T5)
```

This highlights a potential problem with using the index value to identify a location in an array: if we change the size of the matrix, by, for example, making a 7 times table, then the index will no longer represent the value in row 3 and column 5. To see which element we have found, we can again use **ind2sub**.

```
[X7, Y7] = ind2sub(size(SevenTT),Ind3T5)
```

We have instead found the value for the product 2×4 which is equal to 8. Now explore what happens if we don't supply any output variable names when we use **ind2sub**. In this case only one answer is given – the function simply returns the index value.

```
ind2sub(size(SevenTT), Ind3T5)
```

6.2.5 Functions with One or More Required Outputs and Optional Outputs

Many built-in functions offer flexibility both in the number of inputs and the number of outputs. A useful example of this, particularly if you are intending to use MATLAB to analyse data, is the **find** function. To explore the various ways in which we can use this function, we first need to add a data set to the workspace. Suppose we have the results from a field study whereby counts were recorded of the number of three distinct species of beetles found in each of four fallen trees inspected:

```
Beetles = [0 1 0 0
           2 0 1 0
           0 0 0 2];
```

The most basic way to use **find** is with one input, an array of data, and one output (which doesn't have to be declared). This will give a list of index values for the locations in which the value in the input array is non-zero:

```
FoundABeetle = find(Beetles)
```

We may prefer to obtain the subscript locations of the non-zero values, in which case we can supply two output values. The row number of each non-zero value will be output to the first variable and the column number to the second variable:

```
[Species, Sample] = find(Beetles);
SubscriptLocations = [Species Sample]
```

Notice the subscript locations are given in the same order as the index (you can check this using **sub2ind**). We can also use the function **find** to return the number of beetles at each subscripted location:

```
[Species, Sample, NumBeetles] = find(Beetles);
HowMany = [Species Sample NumBeetles]
```

If you want the index locations and the number of beetles, then you need to combine the outputs from the function call with one output, with the output from the call to the function with three outputs as follows:

```
HowManyList = [FoundABeetle NumBeetles]
```

The **find** function can also be used to find the location of values which meet some other criteria, for example, find all locations with two or more beetles:

```
[X Y] = find(Beetles >= 2);
MoreThan1Beetle = [X Y]
```

You may have noticed that sometimes we use commas to separate the output variables and sometimes we don't: the commas are optional, and do not change the code. It is up to personal preference whether you include them. There are two optional input values that can be used to refine the function **find**. The first is used to limit the number of elements output which meet the search criteria to a specified number of values. The second is used to reverse the direction of the search, so instead of starting at the top right and working down the columns, the search starts at the bottom left of the array and works backwards up the columns:

```
[R,C,V]= find(Beetles,2); % outputs the first two non-zero values
LocFirst = find(Beetles,2);
FirstTwoNonZeroObs = [LocFirst V]
[a b c] = find(Beetles,2,'last'); % outputs the last two values
LocLast =find(Beetles,2,'last');
LastTwoNonZeroObs = [LocLast c]
```


Explore what happens when you change the number of outputs requested to more than the number of non-zero values.

6.2.6 Applying Functions to Data 'Structures'

A structure array is a flexible data type that can be used to group related data in data containers which are referred to as fields. A structure can contain multiple fields which can be different data types, and each field can contain a different number of elements. This makes it easier to keep track of related data sets and, as shown in this section, to apply functions to related data sets. For example, suppose we have collected data on the catch from five fishing boats on two different days. We can add this data to one structure, which we will call **NetCatch** using the following dot notation:

```
NetCatch.Day1 = [10 3 4 12 8];
NetCatch.Day2 = [6 7 2 2 1];
```

By placing a dot between **NetCatch** and **Day1,** we have automatically created a structure and added a variable **Day1**. We then add another variable to the structure, in this case **Day2**. To find the maximum catch each day, we can use another function with optional inputs and outputs, **max**. There are at least three different approaches that can be used to find the maximum value within each field in the structure:

- **Option 1**

Treat **NetCatch.Day1** and **NetCatch.Day2** like any other array of data, and calculate each maximum separately:

```
max(NetCatch.Day1)
max(NetCatch.Day1)
```

Option 2

Combine NetCatch.Day1 and NetCatch.Day2 into a single array, and apply the maximum function:

```
NetCatchAllDays = [NetCatch.Day1; NetCatch.Day2]
max(NetCatchAllDays)
max(NetCatchAllDays,[],2)
```

The first use of **max** calculated the maximum for each column of data. To extract the row sum rather than the column sum, we added **,[],2** to the function. You may be wondering why we need to include the square brackets in this function when we didn't need them for finding the mean of the rows using the **mean** function. To see why, first explore what happens when you type:

```
MaxVs2 = max(NetCatchAllDays,2)
MaxVs10 = max(NetCatchAllDays,10)
```

The values in **NetCatchAllDays** are compared with the numeric value after the comma, and if the numeric value is greater than the value in **NetCatchAllDays**, then it replaces that value in the output. We can find the maximum catch in each location by finding the maximum value in each position in the array (the arrays must be the same size):

```
MaxEachLoc = max(NetCatch.Day1,NetCatch.Day2)
```

The square brackets in the command, **max(NetCatchAllDays,[],2)**, tell the function that we don't want to compare the array **NetCatchAllDays** with any external value but do want to change the dimensions along which the maximum is calculated so that the row maximums not the column maximums are output.

Option 3

Apply the **max** function to each field in the data structure by using another built-in function **structfun**:

```
structfun(@max,NetCatch)
```

To use the **max** function, we need to include the @. We can use this method to apply a variety of functions to our data:

```
structfun(@std,NetCatch)
structfun(@mean,NetCatch)
```

Suppose we also want to know where in the data set the maximum value(s) are. To do this we can request two outputs:

```
[MaxVal,IndexLoc] = structfun(@max,NetCatch)
```

The **structfun** function can be applied to structures with fields that contain different data types and of different lengths. Suppose we add another field which has the date of each fishing trip:

```
NetCatch.TripDate = [datetime(2018,12,20), datetime(2019,1,12)]
```


If you try running **structfun** without adding any extra options, the command will fail, and you will get an error report:

```
structfun(@mean,NetCatch)
```

Error using structfun
Mismatch in type of outputs, at index 3, output 1 (double versus datetime).
Set 'UniformOutput' to false.

This tells us that the fields in **NetCatch** contain different data types, and if we want to apply the function **mean** to the data set, we need to allow MATLAB to include different data types in the output. The corrected code is:

```
structfun(@mean,NetCatch,'UniformOutput',false)
```

Suggestions for Further Exploration in the MATLAB® Documentation

To find out more about structures and applying functions to structures, explore the examples in the documentation.

```
doc structfun
doc struct
```

6.2.7 Nesting Functions

Functions can be nested to save unnecessary creation of variables. For example, suppose we have an array of data and we want to find the sum of the values on the diagonal of a magic square. We could do this in three steps using the built-in functions **magic**, **diag** and **sum**:

```
MyMagic = magic(5);
DiagValues = diag(MyMagic)
SumDiag = sum(DiagValues)
```

But if we are not interested in the intermediary values, we can also nest the functions:

```
NestSum = sum(diag(magic(5)))
```

The functions are evaluated from the inside out: so first **magic(5)** is evaluated, then the diagonal of this matrix is extracted, and then the sum of the diagonal is calculated. While it is appealing to be able to write your code in one line of data, you should take care to make sure that your code is still readable: nesting too many functions can make it difficult to read and to check for errors.

6.3 Toolboxes

A wide range of toolboxes can be purchased from MathWorks® to expand the core MATLAB program. They cannot be used as stand-alone products. The available toolboxes are divided into those associated with the MATLAB product family and those that are for use with the SIMULINK® product family. MATLAB product family toolboxes consist of a set of specialist functions and in some cases graphical user interfaces. The functions in toolboxes are used in the same way as functions from the core MATLAB program.

You can find a list of all the toolboxes currently available from MathWorks, together with detailed descriptions of each toolbox, by visiting the MathWorks website (▶ www.mathworks.com) and navigating to the **products** page. Additional toolboxes can be found on File Exchange: these are free to download but come with the proviso that they may not have been rigorously tested or documented (see ▶ Chap. 5 for further information on File Exchange).

The built-in function **plot** belongs to the core program – this makes sense as it is likely that most users of MATLAB will want to be able to create basic plots. On the other hand, **anova1**, a function that performs one-way analysis of variance, is a function that belongs to the Statistics and Machine Learning Toolbox™ and can only be accessed and used if you have installed that toolbox. You can check where the **plot** function is by typing:

```
which plot
```

The output will be listed in the Command Window and will look something like:

```
built-in (\Applications\MATLAB_R2018b.app\toolbox\matlab\graph2d\plot)
```

This informs us that the function plot is in the MATLAB toolbox which is further categorized into 2-D graphs. If you have a different version of MATLAB, then instead of

2018b you will see the name of your version in the filepath. If you have the Statistics and Machine Learning Toolbox installed, then typing

```
which anova1
```

will give something like:

built-in (\Applications\MATLAB_R2018b.app\toolbox\stats\stats\anova1.m)

This tells us that the **anova1** function is in the stats toolbox. However, **if you do not have this toolbox installed**, then the output returned will be:


```
'anova 1' not found
```

To see which, if any, of the available toolboxes are installed with your version of MATLAB, type the following into the Command Window:

```
ver
```

If you are using code developed by someone with access to toolboxes which you do not own or have not installed, then you may get an **unknown function or variable** error. To see which toolboxes are required for a script or a function, we can use a rather long command. To explore this function, we first need to write and save a script to test. Copy the following into a standard script file and save it as StatCode.m

```
WeightDiff = [1.7 0.7 -0.4 -1.8 0.3 0.6];
H0mean = 0;
[Accept pvalue] = ttest(WeightDiff,H0mean)
```

Now copy the following two lines of code into a script file or to the Command Window:

```
[fList plist] = matlab.codetools.requiredFilesAndProducts('StatCode');
plist.Name
```

Running these two lines of code (note you do not need to run the script **StatCode.m**) informs us that the script **StatCode** requires access to functions from core MATLAB and from the **Statistics and Machine Learning Toolbox**. If you do not have this toolbox, then trying to run the **StatCode** script will produce an unknown function error. This is because the **ttest** function used in the script can only be accessed if you have the Statistics and Machine Learning Toolbox. For more details see the following documentation:

```
doc matlab.codetools.requiredFilesAndProducts
```

In this book, we have tried to avoid using any functions from specialist toolboxes so that the examples are accessible to all users. If you want to explore whether a toolbox that you don't have a licence for would be useful, you can trial the product free of charge: information on how to do this is given on the individual toolboxes webpages (found by navigating through the MathWorks website).

6.4 How to Find a List of All Functions in the Core MATLAB® Program

A list of all the functions you have access to can be found by navigating to the home page of the documentation files (if you do not already have a **Help** window open, type **doc** into the **Command Window** to open one). Select the icon which says Explore MATLAB.

You should now be presented with three options located at the top of the documentation page: All, Examples and Functions. Select **Functions**.

This will give you a full list of functions available in the core MATLAB installation. You can choose to list them by category or alphabetically. A list of functions available in the toolboxes included in your licence can be obtained in a similar way: select the toolbox name from the contents menu, and from the menu along the top, select the pull down labelled **more** and from these **functions**. Selecting any of the function names will take you to the relevant help file.

A full list of functions can also be found by navigating around the function, ***fx***, pull-down menu which is in the left-hand scroll bar of the `Command Window`.

In ► Chap. 9 you will learn how to write your own function files. Familiarizing yourself with the type and range of functions included with MATLAB is a good investment of your time: you may find that the function you are interested in writing already exists or that you can use one or more built-in functions to reduce the complexity of your code.

6.5 Accessing Function Code

If you are interested in the code for any of the built-in functions, you can use **edit** followed by the function name to open the code in the Editor:

```
edit polarhistogram
```

This can be useful to explore the syntax of a function and may help you to learn how to structure your own functions (see ▶ Chap. 9 for an introduction to how to write your own functions). For some commonly used functions, the code is not shown. Compare the above with:

```
edit pi
```

It is very important that you do not edit built-in functions in any way as these files are stored on your computer hard drive, and if you alter them, you will permanently alter the behaviour of that function in your copy of MATLAB.

Take-Home Messages

Built-in functions are a core component of the MATLAB® package. Functions are available for most standard mathematical processes and for many specialist tasks. Some functions can only be used if your licence includes access to specialist toolboxes. Supplied functions vary in complexity with some functions designed for a single purpose and others designed to adapt to a range of scenarios. Documentation containing a list of possible syntaxes that can be used, together with example live scripts, is available for every built-in function in the MATLAB documentation files.

It is a good idea to try out functions on small data sets before applying them to more complex problems to make sure that you are using them correctly, particularly if you are applying functions across rows or columns of data. Many functions can be applied simultaneously to an array of data which can avoid the need for more complex programming. Functions can be nested, but overuse of nesting can result in code that is difficult to read and adapt. Do not edit built-in functions.

Graphs and Plots

C. R. Webb, M. Domijan, *Introduction to MATLAB® for Biologists*,
Learning Materials in Biosciences, https://doi.org/10.1007/978-3-030-21337-4_7

What You Will Learn in This Chapter

MATLAB® provides a flexible environment for creating publication standard plots which can be easily modified to meet journal requirements and replicated for alternative data sets.

In the first part of this chapter, you will learn how to create plots using the interactive interface – this can be used to explore the available plotting options and to generate figures quickly which can then be saved as MATLAB FIG-files or used to generate a function file.

The second part of the chapter introduces the most commonly used commands required to incorporate plotting into script files. In the final part of the chapter, you will learn how to generate a movie from a sequence of images.

7.1 Creating a Plot in MATLAB®

There are two distinct approaches to creating plots in MATLAB:

- Point-and-click user interface
- Command line instructions

As you gain experience with MATLAB, you are likely to find that script-driven plotting is the most time-efficient approach. However, the point-and-click user interface allows rapid exploration of the plotting options and can be used to generate MATLAB code automatically so that specific plot layouts can be replicated at a future date.

7.2 Point-and-Click Interface

Before you start this section, it is a good idea to start with a clean workspace.

```
close all % close any figures
clear % empty workspace
clc       % clear Command Window
```

Now create a new script file and save it as **`NewPlot.m`**. Add the following code to the script file to generate an independent variable and a dependent variable.

```
%% New data
t = -pi:0.01:pi;
f = exp(-t).*cos(10.*t);
```

Run the above section of code. You should now see two new variables named `t` and `f` in the Workspace browser. Now follow the steps outlined below:

- Step 1: Using the mouse select the `PLOTS` tab on the toolstrip.
- Step 2: Move to the Workspace browser, and use the mouse to select the variable `f`.
- Step 3: In the `PLOTS` menu on the toolstrip, you should now see the variable `f` listed and a row of icons each representing a potential method for plotting `f`. Use the mouse to select the first icon which is labelled `plot`.

You should now have a new figure window `Figure 1` which contains a damped oscillating wave. If you can't see a figure window, check to see if it is hiding behind the MATLAB interface (e.g., by pressing ALT-TAB if you are using Microsoft Windows). You can make the plot bigger by hovering the mouse over any edge or corner of the figure window, clicking and dragging the edge.

Before you proceed any further, it is helpful to dock your figure to the MATLAB interface – recall that you can do this by clicking on the downward arrow in the top right-hand side of the figure window.

You can explore the other plot options for this variable, however, even though MATLAB can draw them for you, most of the plot types are meaningless for the variable we have created. Each time you try a new plot option, MATLAB overwrites the existing plot. If you want to keep each plot, then you need to select `New Figure` from the `PLOTS` toolbar.

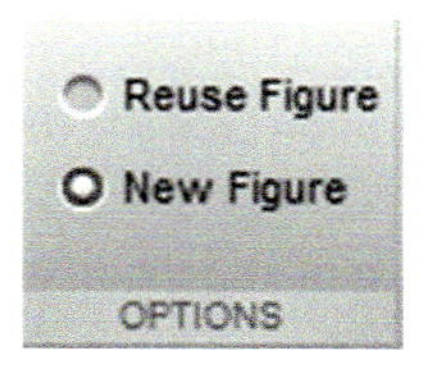

When you have finished creating trying out the options, you can tidy up your desktop by typing:

```
close all % close any figures
```

Click on the icon labelled `plot` to produce the line plot again.

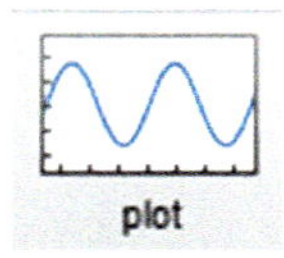

Notice that we did not need to supply the independent variable, `t`, to plot the function `f`. If you look at the x-axis, you will see that the values range from 1 to just over 600. This is clearly wrong as our time range was from $-\pi$ to $+\pi$. When no independent variable is provided, the default is to plot each value in the dependent variable against its position in the vector. There are 629 values in the vector `f` hence the values shown on the x-axis.

To plot `f` as a function of `t`, select both variables in the `Workspace` browser by pressing and holding the CTRL key after you have selected the first variable and then clicking on the next variable. You should now see the following in the `PLOTS` toolbar:

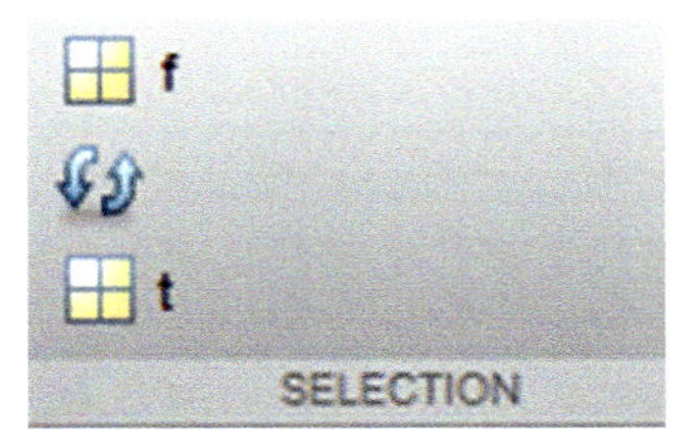

Now click on the `plot` icon. If your variables are in the order shown above, then your plot will be sideways. The first variable listed is assumed to be the independent variable and therefore plotted along the x-axis. To switch the variables around, click on the blue arrows (which you can see in the above screen shot), then click on the plot icon.

Notice there are now some new options for plots. The next option along is **`Plot as multiple series`** (you can see the full name and a help screen by hovering your mouse over the icon). If you click on this icon, a plot with two lines will appear as again MATLAB assumes there is no independent variable and plots both `t` and `f` against the index number of their vector positions.

7.3 Editing Figures Interactively

Use the **`PLOT`** tab on the toolstrip, as described in the previous section, to plot `t` against `f`. For the next task, it may be easier to undock the figure and maximize the figure window. Your plot should look like the plot in ◘ Fig. 7.1.

Along the top of the figure window, there are pull-down menus and a series of icons which allow you to manipulate your figure. Hovering over each icon gives a short description of the icon. To access the properties for the damped cosine wave, find and click on the '`Open Property Inspector`' icon then click anywhere on the plotted line (if you are using an earlier version than 2018b, you can select the **`Show Property Editor`** icon). Explore the Property Inspector window - you can change the line type, colour and width of the plotted line using the pull-down menus. You can also add a marker for the data points.

Now click anywhere on the axes of the figure. The Property Inspector (or Editor in older versions of MATLAB) should now provide a range of options for editing the axes as shown in ◘ Fig. 7.2. These include adding a title to the figure, labelling the axes and changing the font used. These are listed under the subheading Labels.

Clicking on the button labelled **`Ticks`**... allows you to edit the location of ticks along the axes and set their labels.

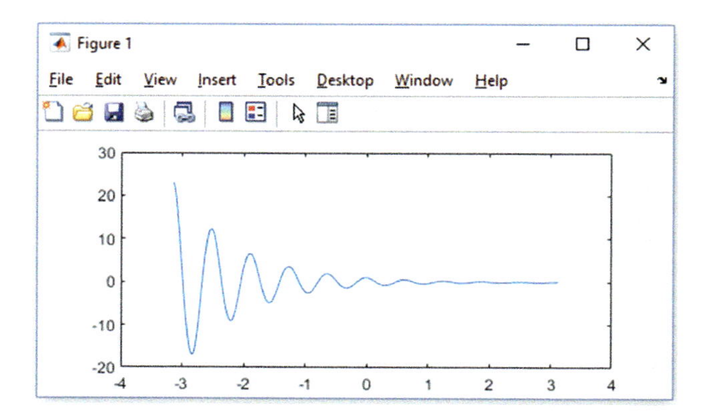

◘ **Fig. 7.1** Plot of a damped cosine wave as displayed in the MATLAB® figure window

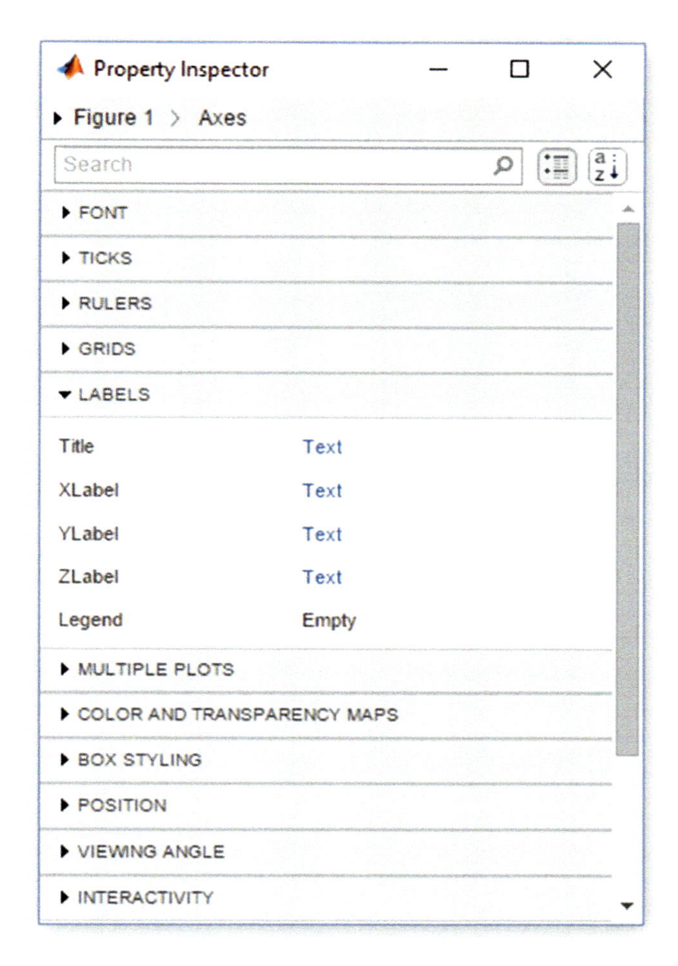

Fig. 7.2 The MATLAB Property Inspector with options for refining the plot axis

Returning to the figure window and selecting '`Insert`' from the plot menu bar reveals options to add arrows and text boxes to annotate your figure.

7.4 Saving a Figure

Once you are satisfied with your figure, you can either send it directly to your printer by selecting `File>Print` from the pull-down menu or use one of three methods to save the figure information:

- MATLAB figure format: `.fig`
- Standard figure formats: `bmp`, `jpg`, `tif` etc.
- MATLAB function file

7.4.1 MATLAB® Figure Format *.fig

The default method for saving figures is as a MATLAB figure file. This is achieved by either using the shortcut CTRL-S or selecting **File > Save**. Figure files can be opened and edited in MATLAB. It is important to note that the data used to generate the figure is not saved explicitly as variables. To see this, save your file as **Test.fig**. Now type the following into the command line to open the figure:

```
close all % close any figures
clear % empty workspace
open Test.fig
```

You should see the figure open in a figure window, but your workspace remains empty. You can now make further edits to the figure as desired, but you have lost the information on the data used to generate the figure.

7.4.2 Figures for Incorporation into Other Documents

To save as one of the standard figure formats such as **bmp**, **jpg** and **tif**, select **File>Save As...** from the menu across the top of the figure window (or in the **FIGURE** tab from the main interface toolstrip if you have docked the figure window). In the pop-up **Save As** box, use the pull-down menu '**Save as type**' to choose a specific file type. A quirk of this feature is that MATLAB will save the figure as it looks on the screen, so if you want the figure to be more stretched along the x-axis, you first need to undock the figure window and use the mouse to get the figure in the shape you want, and then save the figure.

7.4.3 Generate a Function File with Code to Recreate the Figure

This method is useful if you want to be able to recreate the same figure either with the same data or with an alternative data set which contains the same type of variables (they do not have to have the same variable names as you will see below). If you don't have a figure open, repeat the process outlined above to create two variables **f** and **t**, and plot them as a line graph. Edit the graph using the Property Editor to change the line type and colour, and add axes labels. Now select **File > Generate Code** ...

This will generate a function file and open it in the **Editor** window in the main interface. The file will automatically be called something like **Untitled1**. Click on the **EDITOR** tab on the main interface toolstrip, and choose **Save** to save the file: MATLAB will suggest calling the file **createfigure.m**. Accept this suggestion (in ► Chap. 9 we explore creating your own functions in more detail).

Once you have saved the file, type the following into the Command Window

```
close all % close any figures
clear % empty workspace
```

As we have cleared all the variables from the workspace, you will get an error message if you now type **`createfigure(t,f)`** in your Command Window. The function file has saved all the commands to generate the same figure as before, but it has not stored the data points. To use this file, we first need to generate our data again.

```
t = 0:.01:3;
f = exp(-t).*cos(10.*t);
```

Now run the new function and you should get the figure as saved:

```
createfigure(t,f)
```

The advantage of this method is we can generate similar figures for different data. Try the following:

```
x = 0:0.01:pi./2;
g = cos(10.*x);
createfigure(x,g)
```

7.5 Incorporating Plotting Commands into Script Files

The point-and-click method provides a good way to discover the capabilities of the MATLAB plotting environment and can be used to produce your own built-in functions so that you can use a single command to produce a plot to your exact specification for different data sets. If you open the **`createfigure.m`** file (which we created above), you will see that several built-in functions are used to recreate the plot including commands to create a figure and to create axes. These commands will prove useful if you want to create more complex scripts for plotting multiple set of data; however, for now we can get away with a simpler approach.

Each time you plot a new figure using the point-and-click method, you may have noticed that a new command appeared in the Command Window (you can see these by scrolling back through your Command History). This is the command that will enable you to produce the same plot by typing directly into the Command Window or a script file.

Open a new script and add the following commands to it (you should already have these commands in your Command History, so you can save typing by dragging from the Command History to the Editor and modifying some of them so they match the code below).

```
%% plot using built-in functions
close all % close any figures
clear % empty workspace
t = 0:.01:3;
Aphids = exp(-t).*(cos(10.*t).^2);
plot(t,Aphids)
```

The basic plot function requires information on the values on the independent variable (plotted along the horizontal axis) and the dependent variable (plotted along the vertical axis).

7.5.1 Adding a Title and Axis Labels

The commands to annotate the plot are:

```
title('Predicted population dynamics')
xlabel('time, \tau (months)')
ylabel('Number of aphids per m^3')
```

Run the above code, and check to see what happens when we used \tau and m^3. These are LaTeX commands (a typesetting system often favoured by mathematicians and physicists), and they are recognized by MATLAB.

7.5.2 Figure Windows

Recall that the first plot we made using point-and-click method plotted the dependent variable against its location in the vector. The equivalent command is:

```
plot(Aphids)
```

If you docked the plot you created with the command `plot(t,Aphids)` and did not use the `close all` statement before the command `plot(Aphids)`, you probably noticed that the new command cleared the figure window, including the titles and axes labels, and placed your new plot there. If you want to keep the old plot and plot your new plot in a separate window, then you need to define a new figure window using the command `figure`. We will explore this while introducing the use of `plot` on data with multiple dependent variables, and, the function `legend` which adds a key to the plot.

```
%% Method 1: list each x,y pair in plot command
figure
t = -pi:0.01:pi;
f = exp(-t).* cos(10.*t);
g = cos(10.*t);
plot(t,f,t,g)
%% Method 2: concatenate dependent variables
figure(4)
TP = 0:1:10;
MyData = [3.*TP+1; 2.*TP-4; -4.*TP+5];
plot(TP,MyData)
legend('Trial 1','Trial 2','Trial 3')
```

In each case a new figure window is opened. By default, the figures are named sequentially `Figure 1`, `Figure 2` and `Figure 3`. However, supplying an integer input assigns that integer as the figure number. A figure window title can be added by supplying additional inputs:

```
figure('Name','Yeast growth')
t = 0:0.1:10;
Day1 = 10.*(1-exp(-0.1.*t))
Day2 = 10.*(1-exp(-0.05.*t))
plot(t,[Day1; Day2])
```

7.5.3 Customizing Line and Marker Presentation

By default, MATLAB plots each new line with a different colour cycling through a standard set of colours. The quickest method of changing the colour and line types is to use the short codes within the plot command. The various options can be found in the MATLAB documentation:

```
doc LineSpec
```

Try the code below:

```
figure('Name','Slug Trap Data')
ObsTime =2:2:10;
Trap1 = [ 2 10 27 45 46];
plot(ObsTime,Trap1,'c--o')
hold on
Trap2 = [1 8 15 18 50];
plot(ObsTime, Trap2,'r-o',...
   'LineWidth',2, ...
   'MarkerFaceColor','g', ...
   'MarkerEdgeColor','k')
```

In the first plot, we used the short codes to specify a cyan dashed line with a circular marker at the data points. We then introduced the `hold` command. This command retains the current plot while superimposing new plots. Finally, we plotted another set of trap data adding more instructions to refine the plot. The ... syntax (ellipses) is used to inform MATLAB that the code continues onto the next line.

7.5.4 Setting Axis Limits and Tick Labels

If you resize the `'Slug Trap Data'` figure, you will observe that the tick values shown on the axes change to ensure the values on the axes are readable. If you want to force MATLAB to always give the tick labels at specific values, use the function `xticks`:

```
xticks([2 4 6 10])
```


The next set of commands allows you to set axes ranges and rename the tick labels:

```
axis([1 11 0 60]) % xmin, xmax, ymin, ymax
yticks([20 30])
yticklabels({'Plant at risk','Plant dead'})
yline(20)
```

Grid lines can be added in line with tick locations.

```
grid on
```

7.5.5 Using the Dot Notation to Access Figure Properties

Options for refining plots can be accessed by assigning a name to each object that makes up the figure and using the dot notation (this replaces the `get` and `set` commands used in older versions of MATLAB). For example:

```
clf
NSlugs = [50 10 45 61];
SlugPlot = bar(NSlugs) % draws a bar plot
SlugPlot.BarWidth = 0.5
SlugPlot.FaceColor = 'k'
SlugPlot.EdgeColor = 'c'
```

Typing `SlugPlot` into the command window will give a list of properties that can be modified for the `SlugPlot` object handle. Not all properties of the graph can be reached this way. Some properties are attached to the current axes. This can be accessed by assigning a name to the current figure using **gcf** the (get current figure) command:

```
Treatment = {'Control','Wool','Shells','Beer'}
fig = gcf; % get current figure
BarAxis = fig.CurrentAxes;
BarAxis.XTickLabel = Treatment
BarAxis.FontWeight = 'bold'
BarAxis.Color = [0.5 0.3 0.5];
```

In the last line of this code, we used a vector containing three values to define the colour of the background of the figure. These values represent a mixing palette in the order R(red) G(green)B(blue). Try changing the values given in the last line of code to explore this further. If you want to know the RGB code for a very specific shade of magenta, say, then you can use the built-in tool **uisetcolor** to choose a colour from a pallet. This will bring up an interface from which you can choose from a standard set of colours or choose a custom colour. The output from the function is the RGB triplet for your chosen colour.

```
MyFavouriteColor = uisetcolor
```

There are many other options for customizing a figure – often the quickest way to find the code for a specific feature is to edit the graph interactively using the Property Editor in the figure window and then generate a function file to recreate the figure and look through the code in the generated file to find the line that created the new feature you want to include in your script.

7.6 Histograms

To explore the generation of histograms in MATLAB, we will create a set of normally distributed data with mean 60 and standard deviation 15, as shown in the code below. To make the results repeatable, we first initialize the random number generator so that the numbers generated can be generated again. The next two lines of code set a mean and standard deviation for our data. The data is generated using the built-in function **randn**. Once the data is set up, we use the **histogram** command to plot a histogram of the generated data and add a title:

```
close all % close any figures
clear all % empty workspace
rng(1,'twister');
mean = 60;
std = 15;
Pop1 = std.*randn(100,1) + mean;
Hist1 = histogram(Pop1)
title(['\mu= ',num2str(mean) ,...
' \sigma = ',num2str(std)]);
```

Since we gave no indication of how we wanted the data binned, an automatic binning algorithm is used that returns bins with a uniform width, chosen to cover the range of elements in **Pop1** and to reveal the underlying shape of the distribution. The number of bins and bin boundaries can be seen by looking at the object handle we assigned to the histogram plot:

```
Hist1
```

The dot notation can be used both to view existing values and to change them. For example, we might want to check and adjust the upper and lower bounds of the bins and change the number of bins. We have deliberately not included a semi-colon in the first and third line of the code below as we want to be able to see what the current values of **BinLimits** and **NumBins** are:


```
Hist1.BinLimits
Hist1.BinLimits = [0 100];
Hist1.NumBins
Hist1.NumBins = 20;
```

Alternatively, you might want to specify your own bin widths:

```
Hist1.BinEdges = [0 30 40:10:70 100];
```

Multiple sets of data can be displayed on the same axes:

```
hold on
mean = 40;
std = 15;
Pop2 = std.*randn(100,1) + mean;
Hist2 = histogram(Pop2);
```

Suggested Areas for Further Exploration

Explore the differences between functions **histogram** and **histogram2** using MATLAB documentation:

```
doc histogram
doc histogram2
```

7.7 Three-Dimensional Plots

A strength of MATLAB is the possibility to create high-quality three-dimensional plots. The hardest part of creating 3-D plots is in ensuring your data are correctly formatted. Here we consider three types of data that you may want to create a 3-D plot with:

- An array/grid/raster of values representing a landscape
- Two independent variables and a dependent variable which is a function of both independent variables
- One independent variable and two dependent variables which are a function of the independent variable

7.7.1 Plotting an array of values

Many ecological studies and spatial models involve collecting data on a raster. Suppose, for example, you have conducted a 5 by 5 grid transect recording the density of fungi in each location and you want to plot your results. First, enter the data into a script file:

```
Fungi = [0 0.1 0 0 0
    0 0.2 0.3 0.1 0
    0 0.1 0.5 0.2 0
    0 0 0.3 0.3 0
    0 0 0.2 0.1 0];
```

We can plot this data in several ways. First, we can look at it in two dimensions using **imagesc**. This function assigns a different colour to each value in the matrix and scales the **colorbar** based on the range of your data.

```
clf
imagesc(Fungi)
```

A key can be added to show the relationship between the colours and density of fungi. If you do not like the colours chosen by MATLAB, you can use one of the many built-in colormaps either in the command line or by opening the figure Property Inspector and using the search tool to find the Colormap menu.

```
FungiBar = colorbar;
colormap('copper');
```

The colormap consists of rows of RGB values. We can flip the colormap and add tick labels as follows:

```
colormap(flipud(copper));
FungiBar.Ticks = [0 0.2 0.5];
FungiBar.TickLabels = {'No Fungi','Low Density','High Density'};
```

To represent this data in three dimensions, we could plot a 3-D bar chart:

```
bar3(Fungi);
zlabel('Density of fungus');
```

We can alter the tick labels on the axes using the same method we used for the 2-D bar chart:


```
fig = gcf;
BarAxis = fig.CurrentAxes;
BarAxis.XTickLabel = {'A','B','C','D','E'};
```

To explore other plotting possibilities for this data, select the **PLOTS** tab in the main MATLAB interface, and highlight the Fungi variable in the workspace. Not all the plot options will make sense for the Fungi data set as it is relatively small.

You can use the built-in peaks function to create a 49 by 49 array of values which gives you something more interesting with which to explore the 3-D plotting functions:

```
z = peaks;
surf(z);
figure
mesh(z);
```

7.7.2 Plotting a Dependent Variable Which Is a Function of Two Independent Variables

Suppose we have developed a model for some biological system which depends on two independent variables and we want a plot which shows all the possible outcomes for a given range of the independent variables. Unlike the last example, we may not have a grid of all the values, just two ranges of values and a function which gives the value of the unknown variable for any pair of the known values. To plot this type of data, we need to use the MATLAB function **meshgrid**. If you want to understand how **meshgrid** works, try investigating using a small range of values as below:

```
var1 = 1:3;
var2 = 7:9;
[V1,V2] = meshgrid(var1,var2);
CombineVars = V1+V2
```

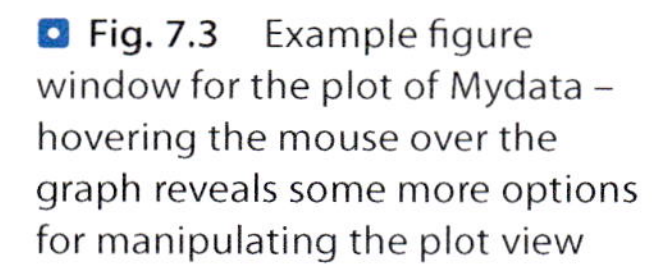

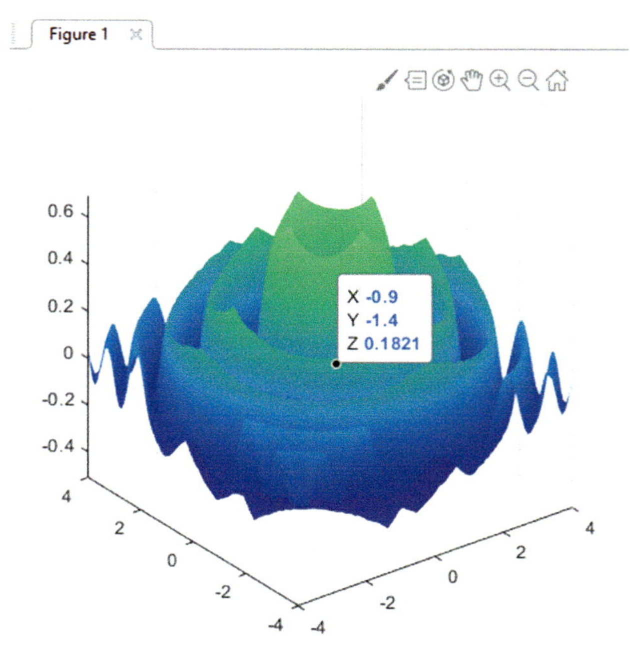

Fig. 7.3 Example figure window for the plot of Mydata – hovering the mouse over the graph reveals some more options for manipulating the plot view

Now use this technique to investigate a more complex function and create a surface plot:

```
x = -4:0.1:4;
y = -4:0.1:4;
[X,Y] = meshgrid(x,y);
MyData = sin(X.^2+Y.^2).*exp(-0.3*(abs(X)+abs(Y)));
colormap(winter);
surf(x,y,MyData);
```

You can get rid of the mesh on the plot by using the built-in function `shading`:

```
shading interp
```

If your copy of MATLAB is recent (version 2018b onwards), then hovering the mouse over the plot window reveals a set of symbols in the top right of the plot window (Fig. 7.3).

These enable you to rotate the plot and zoom in or out. Clicking the home button (a little house) takes you back to the original layout. Alternatively, if you are using an earlier version of MATLAB, you can turn on 3-D rotation by selecting it from the pull-down menu labelled `Tools` (this can be found at the top of the plot window if is not docked or in the `FIGURE` tab of the main MATLAB interface if the plot is docked).

7.7.3 Plotting Two Dependent Variables Which Are a Function of One Independent Variable

Suppose we have collected data on two interacting populations over a series of time points. One way we could plot this data is as separate functions on a 2-D plot (here we have assumed that our data can be described by damped oscillating functions):

```
t = 0:pi/50:10*pi;
st = sin(t).*exp(-.1.*t);
ct = cos(t).*exp(-.1.*t);
plot(t,st,t,ct)
legend('Species A','Species B')
```


Alternatively, we can use a 3-D plot to see the interaction between the two species:

```
figure
plot3(st,ct,t)
xlabel('Species A')
ylabel('Species B')
zlabel('t')
```

Select **Tools>Rotate3d** from the **Figure** menu. Now if you right click anywhere on the graph, a pull-down menu appears from which you can select **Go to X-Y view**. This gives a phase-plane showing the relationship between species A and species B.

7.8 Subplots: Multiple Plots in One Figure Window

The **subplot** command can be used to add multiple plots to one figure window. The basic use of the command requires three inputs: the number of rows into which you want to divide your figure, the number of columns, and where in the grid you want the current figure to go. The position numbers are counted across the rows as illustrated by the next bit of code, which also introduces the plotting of a network derived from an adjacency matrix using the built-in function **graph**:

```
clear
close all
for plotNumber = 1:6
      subplot(2,3,plotNumber)
      Nodes = plotNumber;
      ProbEdge = 0.7;
      AdjMatrix = binornd(1,ProbEdge,Nodes);
      G = graph(AdjMatrix,'upper','omitselfloops')
      plot(G)
      title(['Figure Number ',num2str(plotNumber)])
end
```

In the above piece of code, we used a **for** loop to cycle though plot positions one to six. Loops are covered in more detail in ► Chap. 11. The **num2str** function is used to convert the plot number to a string so that it can be concatenated to the **'Figure number'** text.

For single figure plot numbers, you don't need to include the commas – add the following code to the end of the above script. This code includes an example of how to create a network diagram from an edge list:

```
subplot(234)
% create a graph from an edge list
G = graph({'G1','G3','G4','G4'}, ...
          {'G2','G1','G3','G1'});
plot(G)
title('LOOK AT ME!')
```

The subplot command can be used to divide the window into different sized plots. The first approach is to combine subplots to create larger plots:

```
clear
clf
t = 0:0.1:8.*pi;
subplot(231)
plot(t,sin(t))
subplot(2,3,[2 3])
plot(t,[exp(-0.1.*t); -exp(-0.1.*t)])
subplot(2,3,[4,5,6])
plot(t,sin(t).*exp(-0.1.*t))
```

Alternatively, you can subdivide the figure window into different grid sizes.

```
clf
subplot(511)
bar(randi(10,5,3))
subplot(212)
bar(randi(10,5,3))
```

If the new axes overlap with an existing axis, then the original axes are cleared from the figure window.

The position of subplots can be altered manually in the figure window. To do this first, select the figure window and, if docked, the **FIGURE** tab from the MATLAB interface. Clicking on the arrow symbol allows you to edit the figure. You can now resize any plots within the figure window, by clicking with the mouse and dragging on the axis of each subplot, to get the desired layout. Once you are satisfied with the layout, right click anywhere on the axes of the plot you have resized. This will open a pull-down menu from which you can select **Show code**. This generates a script which contains the position information for your subplot. You can either save this code as a function and use the function as described earlier or you can copy the position into your own script and use the position within a subplot command. Try the following example:

```
subplot('Position',[0.2 0.3 0.8 0.4])
bar3(randi(20,5,5))
```

Suggested Areas for Further Exploration

Explore the functions **subplot** and **graph** in the documentation files:

```
doc subplot
doc graph
```


7.9 Animations

The following section of code uses the built-in function **animatedline** to create a simple movie. The axis command ensures that all plots have the same range of values for the horizontal and vertical axis. We then loop through the values in the function starting at the first time point and use the function **addpoints** to add them to our line and **drawnow** to immediately plot the points in the figure window.

```
clf
clearvars
MyLine = animatedline;
t = -pi:0.01:pi/2;
x=exp(-t).*cos(10.*t);
y = exp(-t).*sin(10.*t);
axis([min(x(:)), max(x(:)), min(y(:)), max(y(:))])
title('Animated line')
xlabel('time')
for k = 1:length(t)
   addpoints(MyLine,x(k),y(k));
   drawnow
end
```

This code uses **for** loops, which we will cover in more detail in ► Chap. 11. If you want to be able to play this animation again, you need to create a movie. To do this add in two commands near the end of the script: **getframe** and **movie**. First, let's create another line, this time choosing a new colour and thicker line width, and save the individual frames (indexed by variable **k**) in the variable **ALMovie**:

```
close all
clear
MyLine = animatedline('Color','m','LineWidth',4);
t = -pi:0.01:pi/2;
x=exp(-t).*cos(10.*t);
y = exp(-t).*sin(10.*t);
axis([min(x(:)), max(x(:)), min(y(:)), max(y(:))])
title('Animated line')
xlabel('time')
for k = 1:length(t)
   addpoints(MyLine,x(k),y(k));
   drawnow
   ALMovie(k) = getframe(gcf);
end
```

The **getframe** function captures the current axes as it appears on the screen as a movie frame. We can replay the frames using the **movie** command:

```
figure
movie(gcf,ALMovie)
```

Note that you can add any type of image to a movie. Try the following example:

```
clf
clear all
for i = 1:50
   [x,y,z] = sphere(i);
   surf(x,y,z)
   SphereMovie(i) = getframe(gcf);
end
clf
movie(gcf,SphereMovie)
```

If you can't see the figure window, remember to check to see whether it is undocked and so hidden behind the MATLAB interface window. Our final example creates a movie of an infection process across a landscape. Copy the code into a script file and run it multiple times:

```
clear all
clf
F = 50; % number of movie frames
W = 100; % width of raster
T = W.*W; % total number of cells in raster
MapI = zeros(W);
MapI(randi([1 T],3,1)) = 1; % initialise disease
beta = 0.1;
```

```
for i = 1:F
    % the next four lines of code find the number of
    % neighbours of each cell that are infected
    kernel = ones(3,3);
    kernel(5) = 0;
    MapIPad = padarray(MapI,[1,1],'symmetric');
    N_Infected = conv2(MapIPad,kernel,'valid');
    % the probability a cell becomes infected is a
    % function of the number of its neighbours that are
    % infected
    ProbInfection = 1-exp(-beta.*N_Infected);
    NewI = binornd(1,ProbInfection);
    MapI = max(MapI,NewI);
    imagesc(MapI)
    M(i) = getframe(gcf);
end
clf
movie(gcf,M)
```


■■ Suggested Areas for Further Exploration

For further information on creating animations and videos, search in the MATLAB documentation for **`Trace Marker Along a Line`**, **`pause`** and **`VideoWriter`**:

```
doc Trace Marker Along a Line
doc pause
doc VideoWriter
```

Take-Home Message

In this chapter we have introduced the basic features of MATLAB® graphics that we have found useful in our research. There are many more refinements that can be included to the plot types we have presented and many more different plot types available. If you are trying to plot a specific type of graph and you cannot find it in the MATLAB documentation files, then try searching in File Exchange on the MathWorks® website (see ▶ Chap. 5). MATLAB can be used to produce publication standard plots which can be precisely reproduced and modified as required.

Anonymous Functions

C. R. Webb, M. Domijan, *Introduction to MATLAB® for Biologists*,
Learning Materials in Biosciences, https://doi.org/10.1007/978-3-030-21337-4_8

What You Will Learn in This Chapter

MATLAB® provides an enormous library of built-in functions; however, there will be times when you cannot find a suitable function. There are two approaches to developing your own functions: anonymous functions and separate function files. Anonymous functions are one-line functions which can be incorporated into a script file and used within that script. They enable you to reuse a line of code multiple times within a script by calling the function. This makes the code easier to read and eliminates the risk of forgetting to change each occurrence of the code if you decide that it requires modification.

In this chapter you will learn how to write and use your own anonymous functions. The chapter starts by considering functions with a fixed number of inputs and then introduces functions which can take a variable number of inputs.

8.1 What Is an Anonymous Function?

An anonymous function is a type of variable that you can include in a script file. They allow you to define an equation with one or more unknown parameter values which can then be used throughout the script without having to rewrite the equation. The main advantages of anonymous functions are:

- Save time by avoiding retyping the same equation in multiple places in a script.
- Reduce the risk of errors by only having the equation in one place.
- Make it easier to modify the equation, and eliminate the chance that you might forget all occurrences of the equation.

Anonymous functions behave like any other variables in a script as they can only be used if they have already been declared in the script or if they are currently in your workspace. If an anonymous function is not in your workspace, then other scripts or functions that do not contain the code for this anonymous function will not be able to use it. If you need to be able to see a function from other scripts, you either need to include the anonymous function in those scripts or create a separate function file (see ▶ Chap. 9).

Anonymous functions belong to the data type called `function handles`. You can read more about this data type in the MATLAB documentation:

```
doc function handle
```

The general structure of an anonymous function is:

```
FunctionName = @(user inputs) equation
```

The `@` operator acts to create a handle. Round brackets after the `@` contain a list of the input arguments – they can be left empty or contain one or more user inputs.

8.2 Creating an Anonymous Function with One Input

The easiest way to understand anonymous functions is to write some. We start with a function which finds the cube of any supplied value. We need to give the function a name: we will call it cube. The function has one user input, the value we want to find the cube of. You can type the following code into the Command Window or a script file.

```
cube = @(x) x.^3
```

Run the code. Look in your Workspace browser – you should see **cube** listed as a variable. It has the value **@(x) x.^3**, and if you hover your mouse over the symbol to the left of the word cube, you will see a pop-up that tells you that **cube** is a 1×1 function handle.

You can now use the function to find the cube of any value. It is a good idea, if possible, to start with a value that you know the answer to. This serves as a basic check that your function behaves as expected.

```
cube(2) % the answer should be 8
```

As with MATLAB built-in functions, you can assign the result to another variable:

```
Eight=cube(2) % the answer should be 8
```

You can also pass a vector or, in the case of this simple function, a matrix to find the cube of several values at once.

```
C = cube([1:4;0.1:0.1:0.4])
```

If you now type **clear** into the Command Window and do not run your script again, you will no longer be able to use the **cube** anonymous function, and an error will be generated.

```
clear
cube(3)
```

8.3 Creating an Anonymous Function with Two or More Inputs

An anonymous function can have multiple inputs. Suppose, for example, we want to write an anonymous function which outputs the volume of a cylinder for any value of the radius (r) and length of the cylinder (L). In a new cell or script type:

```
volumeCylinder = @(r,L) pi.*r.^2.*L;
```

Run this line of script so that **`volumeCylinder`** appears in the workspace. You can now use **`volumeCylinder`** in the same way as a built-in function:

```
% input parameters directly
V1 = volumeCylinder(0.5,15)
% define parameters first
radius = 0.5;
Length = 15;
V2 = volumeCylinder(radius,Length)
% mixture of numerical values and variables
V3 = volumeCylinder(0.5,Length)
```


The two input arguments (r and L) used when the volumeCylinder anonymous function was created are internal to the function. They cannot be seen or accessed outside the function. Outside the function MATLAB doesn't recognize the names – it just expects the correct number of inputs and uses them in the order that they are defined in the inputs to the function. This applies even if you label the inputs with the same names as you used in the function. To see this, compare the output you obtain for V4 and V5:

```
L = 15; r=0.5;
V4 = volumeCylinder(L,r)
V5 = volumeCylinder(r,L)
```

8.4 Anonymous Function with No Inputs

An anonymous function can be a useful way to store command that you need to use repeatedly through the script even if the command does not require any input variables.

```
plotme = @() histogram(randn(30),'EdgeColor','g','FaceColor','m')
```

As **`plotme`** is a function handle, it needs to be called with empty brackets otherwise MATLAB will just repeat the function back to you.

```
plotme()
```

Note that every time you run this function, you will get a slightly different histogram. This is because we are plotting the histogram of 900 values randomly drawn from a standard normal distribution, and each time we execute the function, we will draw 900 different numbers unless we fix the starting seed for the random number generator.

8.5 Fixing Parameter Values in a Function

There may be instances when you want to be able to fix the values of some parameters used in an anonymous function so that it uses the same values whenever you call the function.

For example, suppose you have the results of a generalized linear model in which the number of insects trapped is the response variable and the predictors are light intensity and whether the trap is black or yellow. You may want to develop a script which can be used to make various model predictions, but you are aware that the model fit may change as more data comes in. To do this you must first define the fixed parameter values and then define the anonymous function:

```
%% fixed parameters
clear all
b0 = 0.52;
b1 = 1.54;
b2 = -0.43;
% Col: trap colour code Black = 1 Yellow = 0
InsectNum = @(Watts,Col) b0+b1.*Watts+b2.*Col;
```

The function can now to calculate the expected number of insects trapped for different light levels (in Watts) and by trap type.

```
InsectNum(200,1)
InsectNum(150,0)
```

The parameter values `b0`, `b1` and `b2` are now fixed in the anonymous function until you clear the function from the workspace. If later down the code we define new values of these three parameters, they will not change in the anonymous function unless you rerun the line of code which defined the anonymous function.

```
b0 = 0;
b1 = 0;
b2 = 0;
InsectNum(200,1)
InsectNum(150,0)
InsectNum = @(Watts,Col) b0+b1.*Watts+b2.*Col;
InsectNum(200,1)
InsectNum(150,0)
```

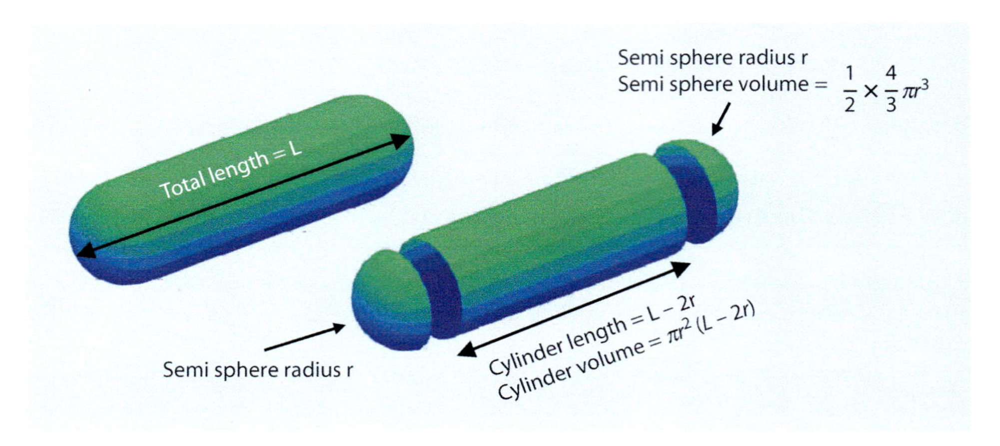

Fig. 8.1 Schematic diagram to illustrate the calculation of the volume of a rod-shaped bacterium

8.6 Using Functions Within Anonymous Functions

If you have a complicated equation, you may want to break it into smaller parts which can be tested and validated which you can then substitute into the main equation. Suppose, for example, that we want to calculate the volume of a set of *Bacillus* bacteria for which we have measurements for the radius and total length. The bacteria can be modelled as cylinder with a hemisphere attached to either end (Fig. 8.1).

One approach to writing an anonymous function to describe this is to reuse the **cube** and **volumeCylinder** anonymous functions created earlier in the chapter. If you have cleared your workspace, you will need to rerun those bits of script or start a fresh script as below:

```
clear
cube = @(x) x.^3;
volumeCylinder = @(r,L) pi.*(r.^2).*L;
volBacillus = @(r,L) volumeCylinder(r,L-2.*r) ...
 + (4/3).*pi.*cube(r);
volBacillus(0.2,5)
```

Suppose next we want to predict how the volume of bacteria in a solution increases over time assuming:

- Bacteria divide in two instantaneously at the same time every hour;
- Each new (child) bacterium has the same volume as the parent bacterium at the end of the hour;
- At time 0 (the time of the start of the experiment) there is one bacterium.

The population volume at the discrete time steps, given by `j`, can now be added to the script:

```
popVolume = @(j,r,L) (2.^j).*volBacillus(r,L);
```

Try out the function with increasing values of `j` – the volume should double at each time step:

```
radius = 0.25;
length = 2;
popVolume(0,radius,length) % volume of 1 bacterium
popVolume(1,radius,length) % volume of 2 bacteria
```

The anonymous function can also take vector inputs enabling quick calculation of the volume of the bacteria over a range of time steps and with differing radii and lengths of the first bacterium:

```
timeSteps = 1:10;
R = [0.2 0.4 0.5]; %radii of bacteria A,B and C
L = [4 6 10]; %lengths of A,B and C
volBactA = popVolume(timeSteps,R(1),L(1));
volBactB = popVolume(timeSteps,R(2),L(2));
volBactC = popVolume(timeSteps,R(3),L(3));
plot(timeSteps,[volBactA; volBactB; volBactC],'-o')
legend('Bacteria A','Bacteria B','Bacteria C')
xlabel('time step')
ylabel('Volume of bacteria (\mu m^3)')
```

The function can take vector inputs as we used element-by-element multiplication in the definition of the popVolume. If you have missed out any of the dots before the multiplication and power signs in your function definitions, you are liable to get an error as MATLAB will attempt to multiply using linear algebra. To see this, try the following code which should generate an error and a suggested solution:

```
popVolume2 = @(j,r,L) (2^j)*volBacillus(r,L);
popVolume2(1:10,0.2,0.5)
```

8.7 Using Anonymous Functions as Inputs to Other Functions

A subset of built-in MATLAB functions accept, or even require, anonymous functions as input parameters. Here we illustrate three examples: `fplot`, `fzero` and `integral`.

Suppose that after an intravenous injection of a clinical drug, the concentration in the blood (in μmg/mol) decays exponentially with a half-life of 2.5 hours. This can be represented as an `anonymous` function with time since injection, `t`, as an input and some fixed value of `C0` (initial concentration of drug in the blood):

```
C0=30; %an arbitrary concentration injected
PlasmaConc = @(t)(C0.*(0.5).^(t/2.5));
PlasmaConc(0:2.5:25) %concentration every 2.5h after injection.
```

We can plot the plasma concentration over first 48 hours, by using the anonymous function as the function handle in the built-in function **fplot**:

```
fplot(PlasmaConc, [0 48], '-o')
```

If we want to know at what time the concentration will drop below 10 μmg/mol, we can use another built-in function **fzero**. This function finds the point(s) at which the plasma concentration **PlasmaConc** decreases below zero. This never happens for our model, and in any case, we want to know when the concentration drops below 10 μmg/mol. A simple fix for this is to modify our anonymous function so that we shift the predicted concentration down by 10 μmg/mol. We can now find out the time at which the blood concentration drops below the critical level. Function **fzero** can take an anonymous function as an input:


```
t0=0;
Crit = 10;
ModifiedConc=@(t)(C0.*(1/2).^(t/2.5))-Crit;
fzero(ModifiedConc,t0)
```

This should tell you that the concentration will drop to 10 μmg/mol at 3.9624 hours after the injection. The second input to the **fzero** function gave the initial value of **t** for the anonymous function.

The exposure of the body to the drug after administration can be expressed as the area under the curve (AUC) which is the definite integral of the plasma drug concentration curve in time. We can compute the AUC using the built-in **integral** function with the plasma drug concentration curve as our anonymous function input. The AUC for curve over the period of first 24 hours is:

```
AUC= integral(PlasmaConc, 0, 24)
```

You can explore the AUC over shorter or longer time ranges by changing the last two arguments (these are the time range over which the area is calculated).

▪▪ Suggested MATLAB® Documentation Files

What happens if you want to change the initial concentration (**C0**)? If you are interested in how you can pass different parameter values to a function such as **integral** where you are evaluating over a range of values, read the section on **Multiple Anonymous Functions** in the MATLAB documentation files:

```
doc Anonymous Functions
```

8.8 Arrays of Anonymous Functions

Anonymous functions can only contain one executable statement; however, we can group function handles together in a cell array (see ▶ Chap. 10 for more information on cell arrays). To create a cell array, use curly brackets. It is also important that there are no spaces as spaces may be interpreted as column separators in a cell array. In the example below, we reuse the moth trap example whereby the predicted number of moths depends on how bright the light in the trap is and the colour of the trap (black or yellow).

```
clearvars
MothNumber = {@(b0,b1,Watts)(b0+b1.*Watts);
               @(b0,b1,b2,Watts)(b0+b1.*Watts+b2.*Watts)};
```

To determine the expected number of moths in a yellow trap, we need to call the function in the first position in the cell (using curly brackets) and provide three inputs (in round brackets). To determine the expected number of moths in the black trap, we call the function in the second position in the cell and provide four inputs.

```
b0 = 0.52;
b1 = 1.54;
b2 = -0.43;
WattRange = 50:50:200;
YellowTrap = MothNumber{1}(b0,b1,WattRange)
BlackTrap = MothNumber{2}(b0,b1,b2,WattRange)
```

The difference between the traps can most readily be observed in a line plot:

```
plot(WattRange,YellowTrap,'y-x',WattRange, BlackTrap,'k-x','LineWidth',4)
legend('Yellow trap','Red Trap')
xlabel('Watts')
ylabel('Number of moths')
```

■ ■ Suggested MATLAB® Documentation Files

For more details on anonymous functions and cell arrays, look up the following:

```
doc anonymous functions
doc cell
```

Take-Home Message

Anonymous functions are one-line functions which enable you to avoid multiple repetitions of identical lines of code. Writing the equation in one place means that if you decide the equation requires modification, then you only need to change it once, avoiding the risk of forgetting to change each occurrence. Anonymous functions can be included anywhere in the code but can only be used after they have been declared. Anonymous functions can only be used if they are in the script you are running or in your current workspace. If you want several scripts to be able to access the same function, then it may be more appropriate to create a separate function file. Separate function files are explored in ► Chap. 9.

Creating Separate Function Files

C. R. Webb, M. Domijan, *Introduction to MATLAB® for Biologists*,
Learning Materials in Biosciences, https://doi.org/10.1007/978-3-030-21337-4_9

What You Will Learn in This Chapter

In this chapter you will learn how to write separate function files which can be used to group together a set of commands. One of the major advantages of MATLAB® is the vast number of built-in functions; nevertheless the MATLAB library of functions cannot cover every possible scenario. Anonymous functions fill some of these gaps; however, separate function files give greater flexibility and can be used in the same way as built-in functions.

Here you will learn about the structure of a function file and work through a range of examples which illustrate how to create your own functions which can accept inputs and produce both numerical and graphical output. You will also find out how to create documentation for your function files.

9.1 Creating a New Function File

There are several options for creating a new function file in MATLAB, and it is up to personal preference which method you use.

- **Option 1**

Select the **HOME** tab on the toolstrip. Select the plus sign icon labelled New, and choose `Function` from the pull-down menu.

9

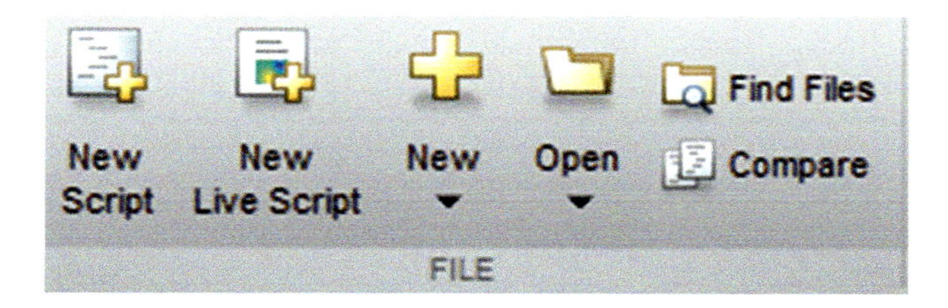

- **Option 2**

Select the **HOME** tab from the toolstrip. From here select the plus sign tab, and choose `Live Function` (only available if your MATLAB version is 2018a or later).

- **Option 3**

Type `edit` into the Command Window. This will open a new script file in the `Editor` window. You can also specify the filename when you use the command `edit`, for example, `edit NewFunc.m` or `edit NewFunc2.mlx`. Using the extension `.m` will create a basic script file, while the extension `.mlx` will create a live script file.

The main advantage of using the toolstrip to open a 'function file' is that the newly opened file includes an outline of the structure of a function. There is no difference between a standard script file and a function file other than what you type into it. The difference between a 'Function' and a 'Live Function' is that you can include more types of explanatory information at the start of a live function file, such as formatted equations, imported images and text formatting.

9.2 The Basic Structure of a Function: Standard Function File

Start by opening a new function file following the first method described above (Option 1). The script should automatically open in the MATLAB Editor and have the following content (if you are using an earlier version of MATLAB, there will only be one output argument and one input argument in the skeleton function file – it is not important as we are going to edit the file):

```
function [outputArg1,outputArg2] = untitled1(inputArg1,inputArg2)
%UNTITLED1 Summary of this function goes here
%   Detailed explanation goes here
outputArg1 = inputArg1;
outputArg2 = inputArg2;
end
```

The file contains the basic framework of a function for a standard script (i.e. not opened using Live Editor). The keyword **`function`** informs the compiler that the following set of commands form part of a function. This must always be matched with an **`end`** statement when you have finished writing the content of the function. The function, currently named **`untitled1`**, accepts the inputs **`inputArg1`** and **`inputArg2`** and returns the outputs **`outputArg1`** and **`outputArg2`**. The body of the function contains a description of the function followed by some calculations to determine the values of the output arguments based on the values of the input arguments.

It is a good idea to keep all the functions you write in one directory. For now, create a new directory in the Current Folder window, and name it **`MyFunctions`**. Make sure the **`MyFunctions`** folder is in the MATLAB search path (see ▶ Chap. 4 if you are unsure how to do this). For our first example, we will create a function which provides the solutions to a quadratic equation. Edit the MATLAB example function so that it has the following content:

```
function [x1, x2] = Quadratic(a,b,c)
% Solves a quadratic equation
% of the form ax^2+bx+c=0
% example: [x1,x2] = Quadratic(1,-1,-6)
% see also sqrt function
PartA = -b./2;
PartB = sqrt(b.^2-4.*a.*c)./2.*a;
x1 = PartA+PartB;
x2 = PartA-PartB;
end
```

Compared to the function template we have:

- Increased the number of inputs to three. A function can have any number of inputs and outputs.

- Changed the names of the inputs to **a, b** and **c**. These are used within the function to calculate our newly named outputs **x1** and **x2**.
- Edited the comments so that they describe the purpose of our function.
- Added a few lines of calculations and created two new variables: **PartA** and **PartB**. These are only visible inside the function.

Save the function to the **MyFunctions** directory: accept the suggested name for the function (which should in this case be **Quadratic**). Notice that the suggested name is the same as the name used for the function in the script. To avoid confusion between the script name and the function name, it is best practice to use the same name for the function and the file.

Before you continue, clear the workspace so there are no variables in the Workspace browser:

```
clearvars
```

The first thing to notice is that by adding comments to the text, we have automatically created a help file for the function. To see this type:


```
doc Quadratic
```

If you included the last line of comments '% see also sqrt function', your documentation will contain clickable links to the documentation for the built-in function **sqrt** and the keyword **function**. Copy the example of the function which we included in the documentation into the Command Window:

```
[Sol1,Sol2] = Quadratic(1,-1,-6)
```

If you have typed the function in correctly, this should give the answers **Sol1 = 3, Sol2 = -2**. Look in the workspace. The only variables in there are **Sol1** and **Sol2**. The variable names used and created in the function, **a, b, c, x1, x2, PartA** and **PartB** are not visible and so cannot be accessed outside the function. The function created here works in the same way as any built-in function (see ▶ Chap. 6). If you call the function without explicitly assigning two outputs, only the first output will be given as you will see if you run either of the following lines of code in the Command Window or from a new script:

```
Quadratic(1,-1,-6)
Solution = Quadratic(1,-1,-6)
```

Suppose we are trying to solve the equation $x^2 - 4 = 0$. We might be tempted to call the function with only two inputs. However, calling the function without supplying enough input arguments will result in an error as all three values are required to evaluate the equations in the function:

```
[ans1,ans2] = Quadratic(1,-4) % produces an error
[f,g] = Quadratic(1,0,-4) % correct
```

9.3 The Basic Structure of a Function: Live Function File

Live function files require MATLAB version 2018a or later. If the version of MATLAB you are using predates this, then you might want to skip to the next section, ***Naming functions***.

Open a new live function file by clicking either the **HOME** tab or **EDITOR** tab on the toolstrip and selecting **Live Function** from the pull-down menu that appears when you click on the yellow plus sign. A live function script will open in the Editor. The content is different to the content we saw when we opened a standard function file (Fig. 9.1).

Just like live scripts (see ▶ Chap. 4), live functions can contain a mixture of code and explanatory text. Only text placed before the code for the function will be included in the documentation file. You can include images, formatted equations and links to other documentation in this section of the file. To explore how live functions work, edit the code part of the script as follows (note that this file includes a conditional statement (**if**) which will be covered in ▶ Chap. 11):

```
function Total = Dispense(dose,freq,duration)
Total = dose.*freq.*duration;
if Total > 500
    disp('You cannot prescribe more than 500ml')
end
end
```

Now click on the **save** symbol in the **LIVE EDITOR** tab. MATLAB will suggest that you save the file as **Dispense.mlx** – accept the suggestion. The **mlx** extension is used for live scripts and functions. Look in the Current Folder browser – each file has a symbol next to it depending on whether they are script files, function files or live versions of each file type.

Fig. 9.1 Default Live function file – the digit after Untitled may be higher if you already have a file labelled Untitled1 open in the Editor

Brief summary of this function.

Detailed explanation of this function.

```
1  function z = Untitled1(x, y)
2  z = x + y;
3  end
```

You can use the new function in the same way as you would use a function created in a standard file or a built-in function:

```
Dispense(15,3,5)
Dose = 30;
TimesADay = 4;
Duration = 7;
Prescription = Dispense(Dose, TimesADay, Duration)
```

The main difference between a live function and a standard function is that you can create more elaborate documentation. Make sure the `Dispense.mlx` function is open in your Editor, and select the window so that you can see the `LIVE EDITOR` tab on the toolstrip. Using the icons on the `LIVE EDITOR` toolbar, you can switch between code and text. Any text that you want to have in the documentation file must be placed *before* the section of code which contains the function.

In text mode you can choose different types of header fonts. For example, when the `LIVE EDITOR` tab is selected, you can use the menus to change a line of font to one of the heading types or underline a section of text. Selecting the `INSERT` tab from the main menu will reveal further options to include additional features such as images and equations.

It is often useful to add to your documentation of the function some example code which is correctly formatted for the user to copy and paste into their Command Window to quickly test out the function. To do this select the `INSERT` tab on the toolstrip, and then select `MATLAB` from the `Code Example` pull-down menu. An example of how you might format the document part of your Live function file is given in ◘ Fig. 9.2.

To see what your documentation looks type:

```
doc Dispense
```

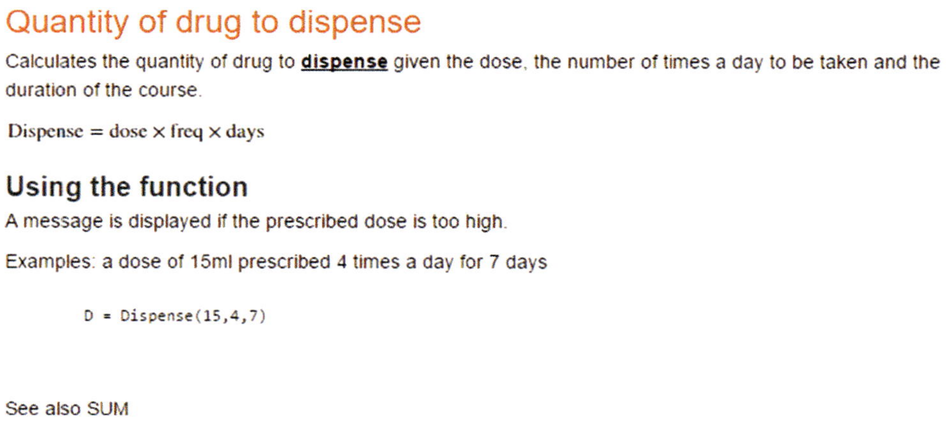

Quantity of drug to dispense

Calculates the quantity of drug to **dispense** given the dose, the number of times a day to be taken and the duration of the course.

$\text{Dispense} = \text{dose} \times \text{freq} \times \text{days}$

Using the function

A message is displayed if the prescribed dose is too high.

Examples: a dose of 15ml prescribed 4 times a day for 7 days

```
D = Dispense(15,4,7)
```

See also SUM

◘ **Fig. 9.2** Example of a descriptive section of **Dispense.mlx**

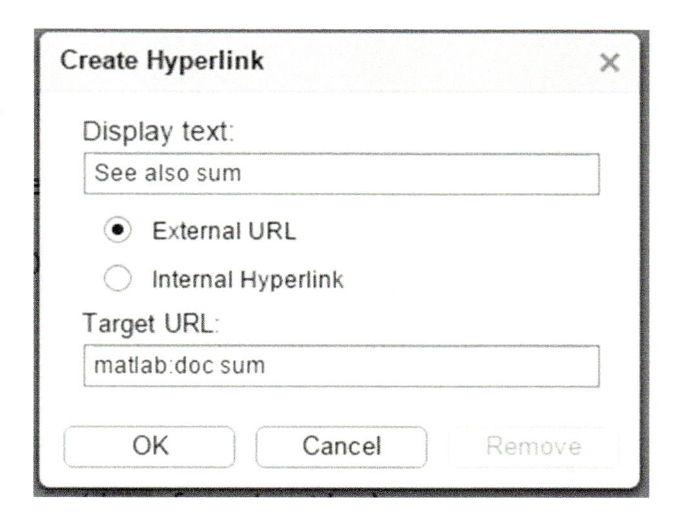

Fig. 9.3 Example of how to fill in the hyperlink pop-up window to give a link to the MATLAB® documentation file for the built-in function **'sum'**

Note that **See Also** links do not produce hyperlinks in the Live Editor (MATLAB 2018b) – this is likely to be corrected in future releases. As an alternative you can add a hyperlink to the text part of your live function. To do this select the **INSERT** tab in the main interface menu, and click on the **hyperlink** icon. In the pop-up that appears, fill in the **Display text** box and the **Target URL** as shown in Fig. 9.3.

9.4 Naming Functions

Before we continue to explore functions, it is important that we discuss the impact of the name you choose for your function.

- The naming conventions for functions are the same as for variables: they must start with a letter, are case sensitive and cannot contain spaces.
- An error message will appear if you try to save your function as one of the MATLAB keywords. You can find a list of these from the Command Window:

```
iskeyword
```

- The function filename should be the same as the name you have used inside the function. Although your function may still work if you choose a different name for the file to the name you have used inside the function, it is considered best practice to use the same name for the function and the file.
- You should try to avoid naming your function the same as a built-in function that you may use in any piece of code you write. To see whether the function name you would like to use already exists as part of the MATLAB library of functions, you can use the built-in function **exist**. The value output by **exist** depends both on whether the name has been used elsewhere and where the name has been used. The function **which** can be used to find exactly where in the search path an existing function is:

```
doc exist
exist sum
which sum
exist Quadratic
which Quadratic
a = 3;
exist a
which a
exist abcd
```

Be careful how you name your functions and script files. If the name you choose belongs to an existing MATLAB function and if the function you have written is in the current search path, then your function will be evaluated first instead of the MATLAB built-in function. You can read more about function order of execution by looking up the documentation on `Function precedence order`. When this error is made, it can be very hard to spot why your code is not working.

9.5 Functions with No Input or Output Arguments

Functions do not need to contain any inputs or explicitly defined output arguments. Copy the following example in to a new script and save the script as **`PauseScript.m`**.

```
function PauseScript
disp(['to continue press space bar'])
pause;
end
```

Similarly, we can include a plot function within a function without declaring it explicitly as an output. Copy the following example into a new script, and save the script as **`NormalPlot.m`**.

```
function NormalPlot
x = randn(1000,1);
histogram(x)
end
```

Next test out the functions in the Command Window.

```
PauseScript
NormalPlot
```

You should see that in both cases something visible happens: in the first case, a message is displayed in the Command Window. In the second case, a new figure is created.

You should suppress output from any lines of code in your function that you don't want output to the Command Window (by placing a semi-colon at the end of the line of code). Edit the **NormalPlot** function to include a variable **y** which gives the mean of the random numbers generated: **do not** put a semi-colon at the end of the new line:

```
function NormalPlot
x = randn(1000,1);
y = mean(x) % no semicolon!
histogram(x)
end
```

Now save **NormalPlot.m** then run the modified function from the Command Window.

```
clearvars
NormalPlot
```

You should notice that as well as plotting the data, the value of **y** is given in the Command Window. However, you cannot use **y** outside the function as it has not been added to the workspace. To check this either look in the Workspace browser or type **y** into the Command Window. If you need to see a variable outside the function, the most straightforward method is to declare it as an output variable (you can also declare it as a global variable; however, this is generally considered bad practice and should be avoided).

```
function y = NormalPlot
x = randn(1000,1);
y = mean(x);
histogram(x)
end
```

The modified function now outputs the mean of the randomly generated numbers and a histogram.

```
NormalPlot % mean is stored in the Workspace in the variable ans
Mu = NormalPlot % mean is stored in the Workspace in the variable Mu
```

9.6 Functions with Text Inputs

The next example is included to demonstrate that inputs to functions do not need to be numbers. Suppose, for example, we have some sequence data and we want to create a matrix which has the value one if there is a match between two base values, and zero if they do not match. Copy the following into a new script file and save as **SeqMatches.m**.

```
function D = SeqMatches(seq1,seq2)
% SeqMatches compares two text strings
% example, type the following in Command Line: SeqMatches('ACT','CTT')
N1 = length(seq1);
N2 = length(seq2);
D = zeros(N1,N2);
S1 = mat2cell(seq1,1,ones(N1,1));
S2 = mat2cell(seq2,1,ones(N2,1));
for i  = 1:N1
    for j = 1:N2
        D(i,j) = strcmp(S1(i),S2(j));
    end
end
imagesc(D)
axes1 = gca;
set(axes1,'XTick',[1:N2],'XTickLabel',S2,'YTick',[1:N1],'YTickLabel',S1);
end
```

The new function uses the built-in function **mat2cell** to break the input text string into individual letters. The **strcmp** function is a logical function used to compare text strings. For more details search the MATLAB help documentation for these functions. The function also uses nested for loops: these will be covered in ▶ Chap. 11.

Test the function out in the Command Window:

```
M = SeqMatches('ACT','CTT')
```

This should generate a 3 by 3 matrix M with the values:

$$M = \begin{pmatrix} 0 & 0 & 0 \\ 1 & 0 & 0 \\ 0 & 1 & 1 \end{pmatrix}$$

and a plot showing the relationship between the values in the strings.

In the above example, we compared two strings of values – each string must be enclosed by single quotation marks, or an error will be returned:

```
SeqMatches(ACT,CTT)
SeqMatches('ACT','CTT')
```

We can assign each string a variable name before using it:

```
Sample1 = 'ACT GTT';
Sample2 = 'ACTGTT';
SeqMatches(Sample1,Sample2)
```

Note that the 'space' in **Sample1** is treated as a separate character by our function. If you try a numerical input, you do not get an error; however, the output is a 1 by 1 matrix:

```
SeqMatches(1234,2244)
```

To understand why this has happened, we can use the built-in command **keyboard** which pauses the function as it is running and allows you to use the keyboard again so you can place inputs from the Command Line. It also allows you to see into the Function Workspace. Open the function file **SeqMatches** in the MATLAB Editor. Now add the following command into a new line just before the final **end** statement of the function:

```
keyboard
```

Save the function and type the following command in the Command Window

```
Match = SeqMatches('ACT','CAT')
```

Two things should now happen:
- The prompt in the Command Window will change to `K>>.`
- The Workspace window will now be titled **Workspace – SeqMatches**. This is the **function Workspace** and contains all the variables which have been created in the function and their values.

Type **S1** at the `K>>` prompt. You will see that the variable consists of three separate elements containing the strings with the letters **A, C** and **T,** respectively. Now quit debugging mode by typing **dbquit** at the `K>>` prompt. Now let's see what happens when we use numbers with no quotes:

```
Match2 = SeqMatches(1234,2244)
```

This will again enter debugging mode and open the Function Workspace window. This time when you type **S1** at the `K>>` prompt, a 1 by 1 cell array will be returned containing a vector with the number 1234. Quit debugging mode by typing **dbquit** at the `K>>` prompt.

Although the numerical sequences do not generate an error, the function **strcmp** will say there is no match between two sequences of numbers not in quotes:

```
strcmp('ABC','ABC')
strcmp(ABC,ABC) % generates an error
strcmp(12,14)
strcmp(12,12)
strcmp('12','12')
```

9.7 Local Functions

Local functions can be used to make a program easier to read by breaking up the code into smaller sections. Local functions can only be seen, and hence called, within the file that contains them. They must be placed after the main function in a function file that uses them.

As an example, we will return to the discrete time logistic model which was introduced in ► Chap. 4. The model is given by $x_{t+1} = rx_t(1 - x_t)$ where x_t is the scaled population size (relative to the carrying capacity of the population) at time step t, and r is the growth rate of the population. The model has received much attention because of the interesting range of dynamic behaviour depending on the precise value of r. For small values, $1 < r < 3$, the model predicts that the population will tend towards a single equilibrium point, as r gets larger the population fluctuates between an increasing number of values. For values in the range $3.57 < r < 4$, the population dynamics are erratic and apparently random. This behaviour is called chaos.

Copy the code below into a standard script file and save the file with the same name as the main function i.e. **LogisticMap.m**. The program contains three functions: a main function and two local functions. The main function is called **LogisticMap**, and it calls the first local function **LogisticEq** twice and the second local function **LabelPlot** once. Calls to local functions from the main function are made in the same way as built-in functions.

To avoid slowing the program down, we have not stored the values of the population at each time point.

```
% LogisticMap(N0,StepSize,DiscardRuns,Num2Plot)
% plots long term solutions of the discrete time logistic model
% N0: initial population density [0 1]
% StepSize: smaller step size for more detailed output (0,3]
% Discard Runs: number of loops of model to run before plotting
% Num2Plot: number of iterations to plot
% Example code: LogisticMap(0.5,0.01,100,20)
function LogisticMap(N0, StepSize,DiscardRuns, Num2Plot)
clf
hold on
NLoops = DiscardRuns+Num2Plot;
for GrowthRate = 1:StepSize:4
    N = N0;
    for loop = 1:DiscardRuns
        NNew = LogisticEq(GrowthRate,N);
        N = NNew;
    end
    for loop = (DiscardRuns+1):NLoops
        NNew = LogisticEq(GrowthRate,N);
        plot(GrowthRate,NNew,'k.')
        N = NNew;
    end
end
LabelPlot()
end
function xtp = LogisticEq(r,xt)
xtp = r.*xt.*(1-xt);
end
```

```
function LabelPlot
xlabel('Growth rate')
ylabel('x')
title('x_{t+1} = r x_{t} (1-x_{t})')
end
```

The basic flow of the code is: for each value of the growth rate, loop through the logistic model until it has reached a cut-off, as specified by the variable `DiscardRuns`, above which we start plotting the density of the population at each time point up to a specified number of time points. To achieve this, we have used a nested `for` loop. Do not worry if you don't understand the details of the code – focus instead on the use of three functions within one function file. Nested for loops are discussed in more detail in ► Chap. 11.

To run the function, select a set of input parameter values, and run from the Command window (remember to make sure that `LogisticMap.m` is in the search path).

```
LogisticMap(0.5,0.01,300,50)
```

There are a variety of things you might like to do to explore this model further: see if you can adapt the function so that the minimum and maximum growth rate can also be input by the function user; explore the impact of the varying the step size, number of discard runs and number of runs to plot on the model output and model run time.

Local functions can also be included in standard script files (from release 2016b onwards), but they must be placed at the end of the script.

■ ■ Suggestions for Further Exploration

To build a function that behaves differently according to the number of user inputs, look at the documentation on how to control program flow using switch statements:

```
doc switch
```

Search File Exchange for `logistic map` to see what is already available to explore the Logistic map.

Take-Home Messages

Building your own functions will enable you to develop and test pieces of code which you can then call from any other script you write. Create a directory dedicated to storing functions you have written from the outset – this will make it easier to find the path that your function is stored in, to check what code you have written in the past and to share your code with others. Add detailed documentation both at the start of your function (including your name and when you wrote it) and comments within the function: it is amazing how quickly most of us forget the logic behind code that was obvious when we were writing it.

Importing and Exporting Data

C. R. Webb, M. Domijan, *Introduction to MATLAB® for Biologists*,
Learning Materials in Biosciences, https://doi.org/10.1007/978-3-030-21337-4_10

What You Will Learn in This Chapter

This chapter introduces methods for importing data into the MATLAB® workspace and for exporting data for use in other applications. You will learn how to import data using the MATLAB interactive Import Tool and how to import data using built-in functions in script files. The chapter introduces the different formats that imported data can be stored in and shows you how to access data once you have imported it. When you are working with experimental data, there is often some data processing required before you can progress with analysis: we will show you how to deal with missing or unknown values in your data set and how to perform some basic statistics. You will also learn a variety of methods for exporting data from MATLAB.

10.1 Creating a Folder for Your Data

We are going to start this chapter by generating some data that can be used to explore methods to import and export data of different formats into and out of MATLAB. We therefore suggest that you start by creating a folder on your computer to store these files. One way to do this is to navigate to the directory you want to place your folder in by using the Current Folder browser in the MATLAB interface and then right-click anywhere in the Current Folder browser to get a pop-up menu from which you can select **New > Folder**.

In some situations, you may want to run a script which generates some data and then save it to a new folder. So instead of the point-and-click method, we will use the command **mkdir** (make directory). First make sure your Current Folder browser and current working directory match. You can do this either by comparing the file path in the address field, which is located below the toolstrip, with the directory shown in the Current Folder browser; or, by typing **pwd** into the Command Window and again comparing with the Current Folder browser. If they do not match, then double-click on the folder in the Current Folder browser to update the current working directory.

10

Now create the new directory using the following code. Because we don't know what folders you already have in your path, the code below first checks to see whether the folder name already exists anywhere in your search path (see ► Chap. 11 for more information on conditional statements). The built-in function **exist** returns a 7 if the folder name is already in use:

```
if exist('Data_folder')~=7 % if Data_folder does not exist
    mkdir Data_folder % make it
end
```

If the folder does not already exist in your search path, then a directory named **Data_folder** should appear in the Current Folder browser. If you happen to already have a folder named **Data_folder**, then think of a new folder name, and use the same method to create it.

Next, we need to make sure that the data you generate in this chapter is stored in the **Data_folder** and that you will be able to access this data. To do this change the working directory to the **Data_folder** using the command **cd**.

```
cd Data_folder
```

An alternative approach would be to create an **addpath** statement which tells MATLAB to include **Data_Folder** in the list of paths in which it searches for documents, and then to include the folder name whenever you are storing data:

```
cd ..
addpath([pwd '\Data_folder']) % replace \ with / if using Mac OS
```

10.2 Generating Data

In this section we will bypass the need to supply data files to the reader and instead start by generating two data sets in MATLAB and show you how to export them as standard data file types. In doing so you will also learn how to use the functions **fopen** to print the data to a comma delimited text file and **xlswrite** to store data in an Excel compatible file.

First make sure you are now working in the **Data_folder**. Check the directory you are in:

```
pwd
```

If you are not in the correct folder, use the Current Folder browser to navigate to it.

10.2.1 Data Set 1: FluData

As you progress through the FluData section, add the provided commands to a script. You can run each section as you add it by selecting the code and pressing F9 or by separating the code into sections using %% and using CTRL-ENTER to run each section.

The data set we are going to generate will consist of hourly body temperature measurements from diagnosis over a 3-day period for three patients recovering from flu. We will replace some of the values generated with missing values. The first step is to generate hourly body temperature measurements for the three patients. Suppose that each patient's temperature decays exponentially from an initial peak to a normal temperature of around 36.5 °C with the rate of recovery varying between the patients. We will also incorporate some random fluctuations in temperature. An efficient way to generate the data is to use an anonymous function (see ▶ Chap. 8) to represent body temperature as we want to use the same equation three times:

```
clearvars
hour=transpose(0:72);
% TAdm: temperature at admission
% rr: recovery rate
% t: time in hours
Bodytemp = @(TAdm,rr,t) 36.5 + ...
    (TAdm-36.5).*exp(-rr.*t')+0.5.*randn(numel(t),1);
hour = 0:72;
Patient1 =Bodytemp(40.5,0.1,hour);
Patient2 = Bodytemp(40.0,0.5,hour);
Patient3 = Bodytemp(41.5,0.05,hour);
PatientData = [Patient1,Patient2,Patient3];
```

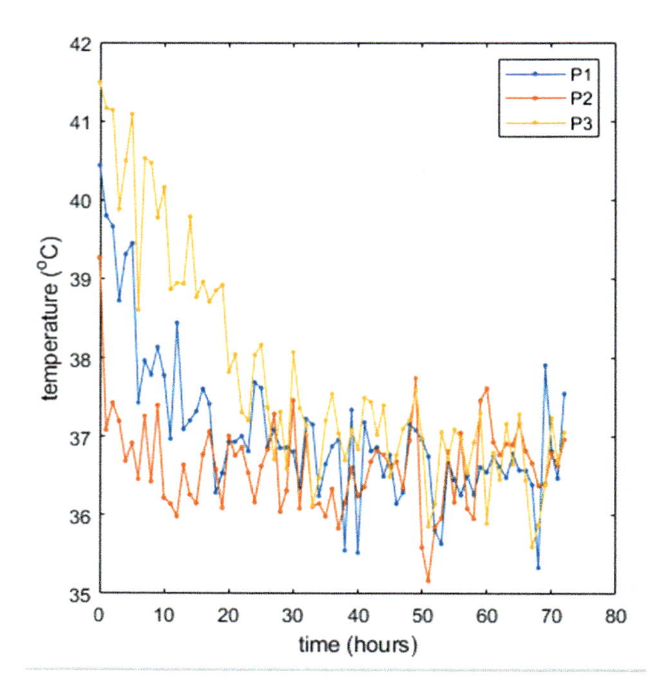

Fig. 10.1 Expected output for the FluData set

If you get an error, make sure you have included all the array operators in the **Bodytemp** equation, used the dot notation (i.e. .* and ./) and included the prime symbol, ', after the t in the exponential function.

Now use the plot function to view the data:

```
clf
plot(hour, PatientData,'.-')
xlabel('time (hours)')
ylabel('temperature (^oC)')
legend('P1', 'P2', 'P3')
```

Your plot should be similar to Fig. 10.1 (it will not be exactly the same as we have incorporated some random fluctuation).

Next remove some of the values and replace them by NaN (not a number: used to represent missing values):

```
PatientData (12,1)=NaN;
PatientData ([25 43 47],2)=NaN;
PatientData ([17 20 34],3)=NaN;
```

Now that we have created a data set, we will export the data to a CSV file. One way to do this is to use the function **fopen**. This function can be used to read data from a file or write data to a file. The first input is the name of the file we want the data to be save to, in this case **FluData.csv**. The second input, **'w'**, specifies the file access type we require: *open or create a new file for writing and discard any contents if it already exists.*

```
FluFile=fopen('FluData.csv', 'w');
```

Now we can use the function **fprintf** to send data to the file. We will start by creating four text labels which will be added to the first row of the file. To do this we use coded characters: **%s** indicates that each input type is a string (i.e. text), and the **\n** indicates that the next **fprintf** statement should print to a new line:

```
fprintf(FluFile,'%s,%s,%s,%s\n','Hour','Patient1', ...
'Patient2', 'Patient3');
```

Next, we will add the data we generated to the file: the hour and temperature data. This time we use the code **%2.0f** and **%4.2f** to store the data in fixed point notation where the digits are used to define the precision at which the data should be stored:

```
fprintf(FluFile,'%2.0f, %4.2f, %4.2f, %4.2f \n', ...
    transpose([hour' PatientData]));
```

The last step is to close the file (notice that for the **fprintf** and **fclose** calls we use the variable name **FluFile** and not the actual name we have called the file, **FluData.csv**):

```
fclose(FluFile)
```

The file **FluData.csv** should now be visible in the Data_folder. If you can't see the file, you may have accidentally saved it elsewhere: you can locate the file using the command **which**.

```
which FluData.csv
```

10.2.2 Data Set 2: qPCR Data

This data represents three replicates of a study in which gene expression is recorded. Measurements are taken every 2 hours over a 24-hour period, and the values recorded are logarithm base 2.

```
clearvars
ObsTime=(0:2:22)';
qPCRout=1.5-3.*sin(ObsTime.*pi./12)+0.5.*randn(12,3);
plot(ObsTime,qPCRout,'.-')
legend('Rep 1', 'Rep 2', 'Rep 3')
```

This time we will use **writetable** to save the data to an Excel file.

One way to prepare the data for export is to (i) create a cell array which contains the titles for each column of data and (ii) convert the **qPCRout** to data class table as shown below:

```
ColLabels={'Observation_Time','rep1', 'rep2','rep3'}
AllDataCell= [ObsTime qPCRout];
T=array2table(AllDataCell, 'VariableNames', ColLabels)
writetable(T, 'qPCRdata.xls')
clearvars
```

Check that the file **qPCRdata.xls** is in the Data_folder.

In this section you have encountered two methods for exporting data using the functions **fopen** and **writetable**. Later in the chapter, we will introduce the functions **csvwrite** and **writematrix**. A list of all the file types that can be imported and exported, together with the relevant functions, can be found by searching the documentation:

```
doc('Supported File Formats for Import and Export')
```


10.3 Data Classes

In this section we present a brief overview of the data classes which are most likely to be relevant to the import and analysis of experimental data. We will start by introducing the three main categories of data that can be stored within arrays. Then we will introduce some alternatives to using standard arrays which can make it easier to conduct data analysis and keep track of variables within a data set.

10.3.1 Numeric, String, Character

While MATLAB does not require you to predeclare variables, each variable is still assigned a data type, and this affects how you can use the variables. The basic data types are:

- Numeric data: These can be stored in different formats. Double-precision floating point is the default format, but values can also be stored as integers or as single precision for efficiency. Infinity is represented as **Inf** and missing values as **NaN**. For more details on how to convert between numeric types, search the documentation for **'Numeric Types'**.
- Nonnumerical values can be represented as character arrays or string arrays. String arrays were introduced in MATLAB 2016b along with a set of functions for manipulating them. The type of quotes you use to encase your array determines the data class:
  - String arrays are created by enclosing values in double quotes (MATLAB version 2017a onwards).
  - Character arrays are created by enclosing values in single quotes.

An important difference between strings and character arrays is that a character array is treated as a set of distinct elements as you will see if you run the following code and look at the output:

```
clearvars
Dog = 'Feed me';
Cat = "Feed me";
Dog(1)
Dog(2)
Dog(5)
Cat(1)
```

The first element in the character array `'Dog'` is the first letter, `F`, while the first (and only) element in the string array `'Cat'` is the whole phrase, `"Feed Me"`.

If we want to extract the second or fifth element from the string which we have labelled `Cat`, we need to use the following notation:

```
Cat{1}(2)
Cat{1}(5)
```

The function `whos` can be used to extract information on the data class and the amount of space required for each variable. You can also display this information in the Workspace browser by selecting `Choose Columns` from the Workspace browser menu.

```
whos
whos Cat
```

In general, string arrays take up more space than character arrays. However, in most cases this difference will not be important. A character array can be converted to a string vector.

```
StrDog = string(Dog)
StrDog(1)
StrDog{1}(1)
```

10.3.2 Storing Data: Cell Arrays, Structures and Table Arrays

In this section we introduce three data classes which can be used to store groups of data:

10.3.2.1 Cell Array

A cell array is an "array of arrays". The elements of a cell array can have different data types and different sizes. There are two main ways to create a cell array. The first method is to assign a name to the cell array using the built-in function `cell` to create a cell array of any size. The example below creates a cell array with three rows and two columns:

```
NewCA = cell(3,2)
```

Elements can be added to individual cells or to groups of cells. Note the use of curly brackets around the input data.

```
NewCA(:,1) = {datetime('yesterday'),datetime('today'),...
    datetime('tomorrow')}
NewCA(:,2) = {10, 20, 30}
```

Elements can be extracted from the array in the usual way but will be of class cell:

```
TodaysDate = NewCA(2,1)
```

To extract numeric data fully from the cell, we need to use curly brackets:

```
NewCA{:,2}
NewCA{2}
```


You can also create a cell array by assigning values to a variable name using the curly bracket notation:

```
InstantCA = {'S1','S2','S3';0.1,0.2,NaN}
```

Converting the elements in a cell array, using the built-in function `string`, to strings enables us to access any character in the array. To extract the first character from the string in the third cell: use curly brackets, to identify the cell number; then round brackets to identify the position in the cell.

```
StrICA = string(InstantCA)
StrICA{3}(1)
```

10.3.2.2 Structure

A structure is similar to a cell array in that it can contain elements of different types and sizes. However, each element is assigned a name rather than a position in an array. Each element in a structure array is referred to as a field. The fields are accessed using dot notation:

```
NewSt = struct('dates',[datetime('today'),datetime('tomorrow')],...
     'name',["Mawr","Bach"], ...
     'weights',[32 33 34])
NewSt.dates
NewSt.weights
```

To access the first element in a field, use round brackets and the element number within the field:

```
NewSt.weights(2)
```

The variable **name** consists of two strings, so we can access the information in this field as described earlier in this section:

```
NewSt.name{1}
NewSt.name{2}(3:4)
```

You can also create structures by using the dot notation without first declaring the structure:

```
InstantSt.timepoints = [1:4]
InstantSt.height = randi(20,4,1)+150
InstantSt.Eyecolor = ["green","blue"]
InstantSt.Eyecolor(1)
```

It is sometimes useful to be able to loop through the values in a field; one way to do this is to combine structures and cell arrays:

```
Trial.Day12 = [2 1 8 4];
Trial.Day13 = [0 2 3 1 1];
Trial.Day14 = [0 3 0];
for TimeStep = {'Day12','Day13','Day14'}
     TimeStep
     Sample = Trial.(TimeStep{1})
     Total = sum(Sample)
     ThirdValue = Trial.(TimeStep{1})(3)
end
```

10.3.2.3 Table Array

Table arrays are the most similar array type to standard spreadsheet data set. Tables store each column of data in a variable. Each variable can contain different data types (e.g. dates, strings, numeric values), but each variable must have the same number of rows. You can import data as a table using either the data Import Tool or the built-in function **readtable**. You can also create tables in MATLAB by combining column vectors:

```
clearvars
Id = transpose(1:10);
timeS = transpose([0:12:108]);
Obs = randi(100,10,1);
NewTable = table(Id,timeS,Obs)
summary(NewTable)
```

Values in a table can be accessed using the dot notation as described for structure arrays:

```
NewTable.Obs
```


Suggestions for Further Exploration in the MATLAB® Documentation

The aim of this section was to give a brief introduction to the data classes which we find useful in our work. Much more detail on each data class and other classes which can be using in MATLAB are available in the documentation files. You may also wish to explore the built-in functions which enable you to convert between different data classes. These can be found by typing:

```
doc('Data Type Conversion')
```

10.4 Importing Data

In this section we will explore some of the ways in which you can import information from external data files into MATLAB. There are two main options for importing data: the interactive option using the data Import Tool and command-driven data import.

10.5 Using the Import Tool App

The MATLAB Import Tool application provides an interactive interface for you to preview data and import all, or a subset, of the data into the workspace. The Import Tool is compatible with most types of spreadsheet files, delimited text files and fixed-width text files. A key feature of the Import Tool is the facility to automatically generate a script or a

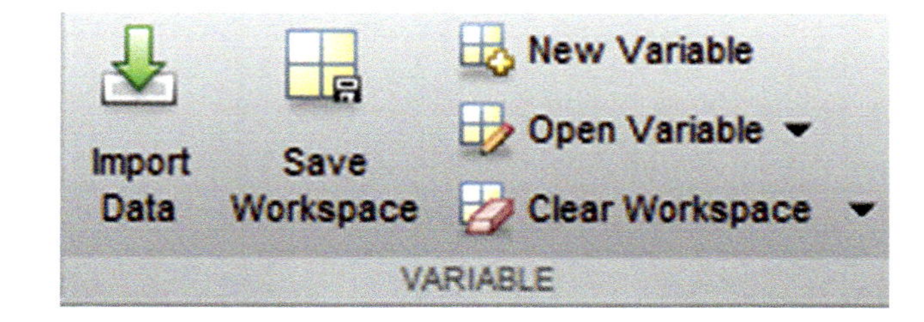

Fig. 10.2 The icon for the Import Data tool can be found in the **VARIABLE** section of the **HOME** tab on the toolstrip

function which is refined according to the choices made in the Import Tool. This enables you to automate data import for future use.

For the rest of this section, we will be working with the fictitious FluData file, `FluData.cvs`. If you have not already created this file, go back to the section on Generating Data which was presented earlier in this chapter and follow the instructions to generate data set 1.

Before you continue clear the workspace:

```
clearvars
```

In the Home tab of the toolstrip, locate the Import Data icon (as shown in Fig. 10.2).

Click on the icon to open a pop-up window entitled `'Import Data'`. Navigate to the folder we created earlier called `Data_folder`, and select the file we created and named `FluData.csv`. The Import Tool will show a preview of the data, with four columns of numeric data and a header row displaying the labels: Day, Patient 1, Patient 2 and Patient 3. If the data looks a little strange, make sure you have selected the, delimited option in the top left of the Import Tool. Your data set should look something like Fig. 10.3 (remember your values will be slightly different as we included some random variation when we generated the data). The data set has a row of headings and 73 rows of data.

At the top of the Data Import window, there are two tabs: Import and View. View allows you to tile multiple data sets if you have more than one data sets open in the Import Tool.

Select the `Import` tab. This contains menus from which you can refine how your data is imported. Below we describe each section of the Import Tool menu:

10.5.1 Delimiters

This section is used to select whether your data is fixed width or delimited and, if it is delimited, what type of delimiter is used. The default is comma delimited which is the delimiter we used when we saved the `FluData`. If you open the `Column Delimiters` menu, you will see other types of delimiters and the option to enter a custom delimiter.

10.5.2 Selection

This section can be used to select a subset of your data set if you don't want to import everything. You can also use the mouse to select a subset of data from the spreadsheet. If

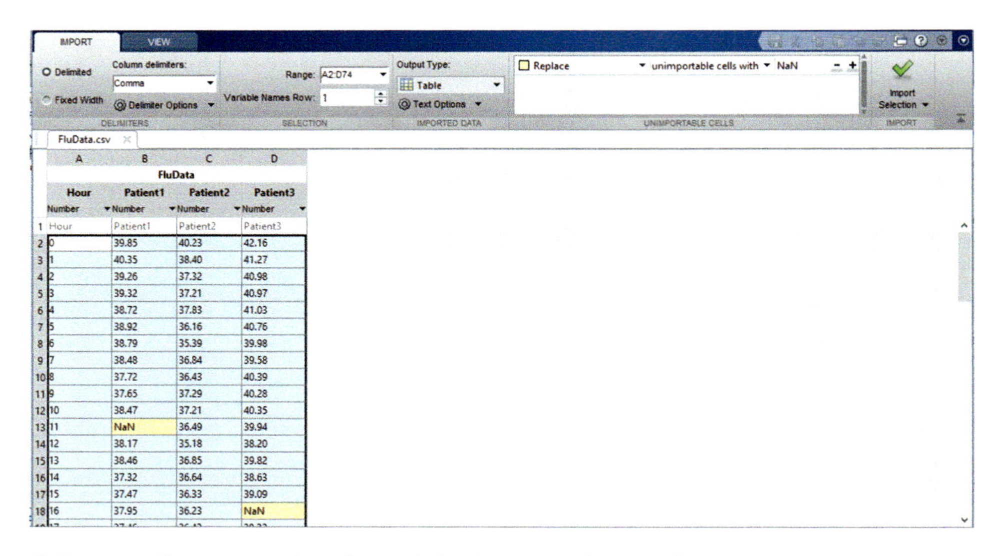

Fig. 10.3 The Import Tool Interface with the FluData.csv. This part of the menu is used to select the data to be imported

you want to skip rows or columns, then after you have selected your first set of data, press and hold the CTRL key on your keyboard. You can also select which row contains variable names, by default this is set to 1. We want to import the entire data set, so set the range to `A2:D74.`

10.5.3 Imported Data

This section is used to select the format you want the data to be stored in when it gets imported into MATLAB. There are five choices:

- Table: imports the data into a table array.
- Column Vectors: imports each column of data into a separate vector and uses the column title to name the vector.
- Numeric Matrix: imports the data into a matrix with the same name as the file you are importing from. If you use this method, you should exclude the title row before you import the data.
- String Array: imports the data into a string array whereby each element in the data set is converted to a string regardless of whether the original data contain text or characters.
- Cell Array: imports the data into a cell array. The first row of the cell array will contain the column headings. Each element in the data set will be assigned to a different cell. This is useful if your data consists of a mixture of numeric and text data.

The choice of method used to import the data depends on how you will be using the data in your code. For now, select `Column Vectors.`

10.5.4 Unimportable Data

This part of the menu is used to select how to import any 'non-importable values' or blank cells. In the preview, any non-importable values will be highlighted in yellow. You can use the menus here to create a set of rules to tell MATLAB whether to import rows of data which include unimportable data for each unimportable data type; and if you decide to import rows containing unimportable data, what the unimportable values should be replaced with. Here we will accept the default setting which is to replace unimportable cells with `NaN` (i.e. in the case of this data set, don't change the unimportable cells).

10.5.5 Import

The final section contains a pull-down menu labelled Import Selection with three distinct options:

- Import Data: imports data in specified format into the MATLAB workspace
- Generate Script or Live Script: creates a script file which can be used to import any data set with the same structure using the specified format into MATLAB
- Generate Function or Live Function: like Generate Script but creates a function rather than a script file

We will explore the second two options later in the chapter. For now, select **Import Data.** You will get a pop-up message informing you that the data has been imported into four vectors (Fig. 10.4).

Minimise the Import Tool window. The four numeric vectors listed in Fig. 10.4 should be visible in the Workspace browser. Notice the vector names are the same as the name in the header for each column of the raw data. You can now work with this data, as with any other MATLAB data. We might start by plotting the data in three subplots:

```
subplot(311)
plot(Hour, Patient1)
title('Patient 1')
xlabel('time (hours)')
ylabel('Temperature ^oC')
% repeat the above code for Patient 2 and 3 so you have three subplots
```

Fig. 10.4 Pop-up message displayed after clicking on **Import Data**

The following variables were imported:
Hour (73x1)
Patient1 (73x1)
Patient2 (73x1)
Patient3 (73x1)

Notice the gaps in the plotted lines: these are where the **NaN** values are in the data sets.

Next, repeat the data import process, but this time select Numeric Matrix as the output type. This time you should get a 73 × 4 matrix named after the import file name, **FluData**. The column labels are not imported, and the information is effectively lost.

The method you use to import the data depends on a combination of the type of information in your data set (are there any nonnumeric data?), what you want to do with the data and personal preference. For many applications the table array output is likely to be the most convenient: table arrays retain information on the variable names; and, each column can be of a different data type.

Now use the Import Tool to import **FluData.csv** with output type **'Table'**. The data range should be set to A2:D74. This will add a 73 × 4 table named **FluData** to the Workspace browser in MATLAB. Notice that the table array **FluData** replaces the array imported when we selected Numeric Matrix.

Open the table in spreadsheet view by double-clicking on the name **FluData** in the Workspace browser. You should see that the columns have preserved the labels from the imported file. Now explore the output from the following commands:

```
istable(FluData) % check FluData has data class type table (1 = true)
summary(FluData) % shows table summary
```


Subsets of the table can be extracted from the table array using row and column numbers, or using row and column titles, or a combination. In this data set we don't have any row titles. If **round brackets** are used, the extracted data are also of data class **tables**.

```
SubT1 = FluData(1:5,:) % first 5 rows in table
SubT2 = FluData(1:5,2) % first 5 rows for column 2 (Patient 1)
SubT3 = FluData(1:5, 'Patient1') % first 5 rows for Patient 1
```

To extract data and change the data class, use **curly brackets**:

```
P1 = FluData{1:5, 'Patient1'}  % extract patient 1 data
```

The function **stackedplot**, introduced in MATLAB 2018b, enables you to plot table array data in one figure window. In addition, hovering the mouse over any part of the line graphs will reveal a data cursor which can be dragged back and forth to compare the temperatures of the patients at any given time point.

```
clf
stackedplot(FluData,{'Patient1','Patient2','Patient3'},...
    'XVariable','Hour','Title','Temperature chart')
```

If you are using an earlier version of MATLAB, you can achieve a similar plot using a **for** loop (**for** loops are presented in detail in ▶ Chap. 11). Notice the use of dot notation as an alternative method to access the data in the table array:

```
i = 1;
for P = {'Patient1','Patient2','Patient3'}
    subplot(3,1,i)
    plot(FluData.Hour,FluData.(P{1}))
    ylabel('Time (hours)')
    title(P)
    i = i+1;
end
xlabel('Temperature ^oC')
```

More information and examples showing how table arrays can be created and manipulated can be found in the MATLAB documentation

```
doc Table
```

10.6 Generating a Script or a Function Using the Import Tool Commands

Attempting to record and relay to a colleague which buttons you clicked-on to import a set of data so that they can replicate previous results can be a frustrating process. However, trying to code a complex data import protocol can be equally frustrating. The Import Tool provides a mechanism by which you can use a point-and-click process to specify how you want your data imported and then automatically generate a script or function which will import your data to these specifications. The final step is knowing how to use these generated files. We go through each file type below.

10.6.1 Import Data Tool: Generate Script

Start by clearing your workspace (e.g. using **`clearvars`**) so you can see the impact of importing the data using Generate Script. Use the Import Tool to open **`FluData.csv`**, remembering to choose comma delimited and select output type **`Table`**. This time select **`Generate script`** from the **`Import Selection`** menu.

You should notice two things: a new script will open in the Editor labelled Untitled1 (or a different number if you have already got some untitled scripts open), and the Workspace browser will remain empty.

Save this script in your **`Data_Folder`** as **`ImportFluData.m`** but don't close it. Now run the script. The data set **`FluData`** will appear in the Workspace browser and is ready for use. You could now include a call to this script in another script, and the data would automatically be imported.

It is worth exploring the script – there are a lot of explanatory comments and relatively few lines of code. Find the line in the code which is preceded by the text `% `**`Import the data.`** This tells the script where to find the data file and what to call the imported table. If you move the original data file to a different directory, share this script, or use the file to import another data set with the same structure; then you will need to modify this path.

The remainder of the file uses commands which are the same as, or similar to, commands you have encountered earlier in this chapter and in the book.

10.6.2 Import Data Tool: Generate Function

Again, start by clearing the workspace so you can see the impact of importing the data using the option `Generate Function`. Use the Import Tool to open `FluData.csv`, remembering to choose comma delimited, and select output type `Table`. This time select `Generate function` from the `Import Selection` menu.

The data are not imported into the Workspace; however, a new function file is opened in the Editor labelled Untitled2 (or a different number if you have already got some untitled scripts open). Edit the first line of the function file replacing the generic function name `importfile` with a more identifiable name `GetFluData`. Save the file as `GetFluData.m`. Do not close the file.

You will get an error if you try to run the file. The function `GetFluData` requires three inputs. Near the top of the initial section of comments, around line 12 of the code, there is an example included in the function file that shows you how to use the function to import the data set which was used to generate the function file. Edit the example to change the function name from `importfile` to `GetFluData`. Now open the documentation file which is automatically generated from the function:


```
doc GetFluData
```

Copy the example into the command line[1]:

```
FluData = GetFluData('FluData.csv', [2, 74]);
```

The data set `FluData` will appear in the Workspace browser and is ready for use. The key difference with the function is that there is some flexibility in how it can be used. Neither the file path nor the filename is specified, and we can choose which rows of data to import. You can now use this function just like any other function as long as it is visible in the search path (see ▸ Chap. 9 for more details on creating and using your own functions).

10.7 Import Data Using the Command Window

MATLAB release 2018b or earlier

Both the automatically generated script and function use the same set of built-in functions to open the file (`fopen`), read in the data (`textscan`) and close it (`fclose`). There are a number of other functions which can be used to import data from a range of

1 Note: for earlier releases of MATLAB the square brackets may not be required.

file types. Two more possibilities for comma delimited data files are **importdata** and **csvread**.

The function **importdata** can be used to import a variety of file types including comma delimited data.

```
clearvars
delimiter = ',';
headerline = 1;
FluData=importdata('FluData.csv',delimiter,headerline)
```

The above code imports **FluData.csv** into a structure array with three fields: **FluData.data** are the numerical values, and **FluData.textdata** and **FluData.colheaders** are the text and column header, respectively.

Some functions are refined to work for a subset of file types. For example, the function **csvread** is designed only to read comma delimited data, so any header text rows have to be excluded by giving the first row and the first column to be imported. In contrast to most indexing in MATLAB, the first row and column are identified as row 0 and column 0. The first row of FluData.csv is a header row so we need to exclude this:

```
clearvars
csvread('FluData.csv',1,0) % filename, start row, start column
```

MATLAB r2019a
The automatically generated code uses **readtable** to import data. The function **csvread** is no longer recommended and has been replaced by **readmatrix**.

10.8 Unknown and Missing Values

By default, MATLAB will replace unknown or missing values by the value **NaN** (which stands for 'not-a-number'). You will have seen this in the Import Tool when you were given an option on how to deal with nonnumeric and missing values.

The presence of **NaN** values in your imported data can affect the output of some of your analysis giving potentially erroneous results. Import **FluData.csv** as a table array. You can either do this using the data Import Tool or use the function file we generated earlier. The code below shows you how to import the data using the function file that was generated earlier in this chapter (see Import Data Tool: Generate Function):

```
clearvars
FluData = GetFluData('FluData.csv', [2, 74]);
```

Now calculate the mean of temperature for each patient:

```
muP1 = mean(FluData.Patient1)
muP2 = mean(FluData.Patient2)
muP3 = mean(FluData.Patient3)
```

Since each patient data set includes at least one missing value, represented by **NaN**, then the means cannot be defined. Thus muP1, muP2 and muP3 are also **NaN**. One option is to ignore the locations in which data are missing and use functions developed specifically for working with data with missing values. These functions exclude NaN values from calculations. Alternatively, we can add an option to the standard functions to tell them to ignore NaN values:

```
% method 1: use special nan function
mu2P1 = nanmean(FluData.Patient1)
std2P1 = nanstd(FluData.Patient1)
% method 2: use standard function but include 'omitnan' if available
mu3P1=mean(FluData.Patient1,'omitnan')
std3P1=std(FluData.Patient1,'omitnan')
```


It is useful to know how many, and where, the **NaN** values appear in your data, so you can decide how to deal with them. You can check whether your data contains any missing values by using the built-in function **isnan**. The output is a logical array with a 1 where the values are NaN and 0 elsewhere:

```
TestForNaN = isnan(FluData.Patient3)
```

The total number of missing values can now be calculated using the function **sum**:

```
NumMissing = sum(TestForNaN(:))
```

This tells us there are three missing values in patient 3's data. Suppose we want to replace missing values with some other value, say the temperature either side of the missing reading. To do this we combine the **find** function with **isnan** to find the location of the missing values and then replace them with the mean of their neighbouring values[2]:

2 In practice this code would need refining to allow for the possibilities that the missing values are at the start or end of the data set or that neighbouring values are also missing values.

```
TestForNaN = isnan(FluData.Patient3);
LocMissing = find(TestForNaN);
FluData.P3Corrected = FluData.Patient3
for n = LocMissing
   FluData. P3Corrected (n) = (FluData. P3Corrected (n-1)+...
        FluData. P3Corrected (n+1))./2;
end
FluData. P3Corrected (LocMissing)
```

Here we are using **for** loops to replace each missing value based on its neighbouring values (**for** loops are covered in ► Chap. 11). However, it transpires that there is already a built-in function, **fillmissing** (introduced in MATLAB release 2016b), which we can use to replace the missing values for a range of assumptions. For example, we could replace each missing value with a fixed value or use linear interpolation as we did using the **for** loop:

```
NaNP2 = sum(isnan(FluData.Patient2)); % check how many NaN
FluData.Patient2=fillmissing(FluData.Patient2,'constant',36);
NewNaNP2 = sum(isnan(FluData.Patient2)); % check again how many NaN
FluData.P3CorrectV2 = FluData.Patient3;
FluData.P3CorrectV2 = fillmissing(FluData.P3CorrectV2,'linear');
% compare results of fillmissing with results of for loop
Original = FluData.Patient3(LocMissing)
Method1 = FluData.P3Corrected(LocMissing)
Method2 = FluData.P3CorrectV2(LocMissing)
CompareValues = [Original Method1 Method2]
```

For further exploration of how to deal with missing data in Tables, search the MATLAB documentation for **Clean Messy and Missing Data in Tables**.

10.9 Importing Data from Microsoft Excel Files

Instructions if you are using MATLAB r2018b or earlier

In this section we will be importing the fictitious gene expression data file, **qPCRdata.xls** (or **qPCRdata.xlsx** if you are running your work on a Macintosh machine). The import functions presented here have been superseded in MATLAB r2019a by the functions **readmatrix** and **readcell**, however we have left this section in for users of older versions of MATLAB and code written pre r2019a. If you have not already created this file, go back to the section on Generating Data which was presented earlier in this chapter, and follow the instructions to generate data set 2.

Before you continue clear the workspace, and make sure your current working directory is the **Data_Folder** in which you stored the file **qPCRdata.xls**[3].

```
clearvars
```

3 For some operating systems the extension maybe **xlsx**.

Before importing the data, if you are using a Microsoft Windows, you can get a preview of the structure of the file using the function **xlsfinfo**. The function can generate three outputs. Here we are interested in the first two which we will assign (i) status and (ii) sheets.

```
[status,sheets]=xlsfinfo('qPCRdata')
```

Status confirms that the file you have selected is compatible with being a 'Microsoft Excel Spreadsheet'. Sheets are a cell array: in this case there is only one cell as our data set only has one sheet, labelled 'Sheet 1'. Next you can use the function **xlsread** to import the data from the file into MATLAB. If you are working on Microsoft Windows, type:

```
[qPCR,titles]=xlsread('qPCRdata.xls',sheets{1});
```

else if you are using macOS or Linux, you should type instead:

```
[qPCR,titles]=xlsread('qPCRdata.xls'); % .xlsx for Mac iOS
```


The imported data is split into two variables: **qPCR** is a 12 by 4 matrix of numeric data, and **titles** is a 1 by 4 cell array containing the column headings.

Instructions if you are using MATLAB r2019a onwards

The newly introduced (r2019a) function **readmatrix** can be used to quickly extract the data from our file as follows:

```
qPCR = readmatrix('qPCRdata')
```

The first column of the **qPCR** matrix contains the time points at which measurements were taken. The remaining columns contain log2 ratios of the relative expression levels of the gene (each column is a different replicate).

Next, we will conduct some basic analysis of the data. First separate out the independent variable (time) from the dependent variables (the three replicates containing qPCR output). The values in the dependent variables are log2 ratios. When we extract the dependent variables from the data, we can also convert them to raw ratios by taking 2 to the power of each value. The underlying matrix structure of MATLAB makes this process easy:

```
PCRvals=2.^qPCR(:,2:end);
ObsTime = qPCR(:,1);
```

Note the use of the function **end**: it replaces the need for supplying the total number of rows which can be useful if you want to use a function or script on a data set with the same fields but different numbers of observations.

Now suppose we want to find average value at each time point across the replicates and calculate the standard error of their means. We calculate the standard error of the means (SEM) by first calculating the standard deviation:

```
meanPCR=mean(PCRvals,2); % mean of three replicates
stdevPCR=std(PCRvals,0,2); % standard deviation of three replicates
semPCR=stdevPCR./sqrt(3); % standard error of the mean
```

Now we can plot the raw data and the mean curve with error bars:

```
subplot(211);
plot(ObsTime,PCRvals,'.-')
xlabel('Time (hours)');
ylabel('Relative Expression Levels');
subplot(212)
errorbar(ObsTime,meanPCR,semPCR,'b', 'LineWidth',2);
xlabel('Time (hours)')
ylabel('Relative Expression Levels')
```

10.10 Exporting Data

As with importing data, there are several methods available to export data to standard file formats. When we created the FluData.csv and qPCRdata.xls, we used **fopen** combined with **fprintf** and **xlswrite**, respectively. If we just want to save all the variables (or a selection of variables) currently in the workspace for future use in MATLAB, then we can save them in a MAT-file using **save**:

```
clearvars
Var1 = magic(4);
Var2 = {'age','id','height'}
save('MyData') % saves to a file called MyData.mat
```

To import a MAT-file into MATLAB, either open it from the Current Folder browser or type:

```
clearvars % clear workspace first so you can check effect of load
load('MyData')
```

The function **csvwrite** can be used to save a matrix to a comma-separated text file:

```
RandomData = randi(100,10,3);
RandomData(2:3,:) = NaN; % replace some of the values with NaN
csvwrite('MyNewData.csv',RandomData)
```

If you are using a newer release of MATLAB (r2019a onwards) then it is recommended that you replace the csvwrite command with:

```
writematrix(RandomData,'MyNewData')
```

Now open the file:

```
clearvars
csvread('MyNewData.csv')
```


If you are using a newer release of MATLAB (r2019a onwards) then it is recommended that you replace the csvread command with:

```
readmatrix('MyNewData')
```

If the data you want to save is a mixture of numeric and nonnumeric data, then one option is to use the function **writetable** to save your data. First, we will create a table array in MATLAB and then save the data:

```
Health = {'S';'I';'R'}
NumAsh = [27;50;10];
NumOak = [3;12;2]
Action = {"leave";"treat";"replace"}
Tree = table(Action,NumAsh,NumOak,'RowNames',Health)
writetable(Tree, 'TreeData', 'WriteRowNames',1)
```

Open the file **TreeData.txt** outside MATLAB to see how the data has been stored. Now clear the workspace, and read the data back in using **readtable**:

```
TreeTable = readtable('TreeData')
```

Suggestions for Further Exploration in the MATLAB® Documentation

There are many different file types that can be imported to and exported from MATLAB. A comprehensive list of supported file formats can be found in the MATLAB documentation:

```
doc 'Supported File Formats for Import and Export'
```

A detailed guide to dealing with data in MATLAB can be found in the documentation:

```
doc 'Data Import and Analysis'
```

Take-Home Message

MATLAB® can import data from a wide variety of file formats. For some file formats, you may need access to specialist toolboxes.

The simplest way to import many data sets for the first time is to use the MATLAB Data Import Tool. This can be used to automatically create a script or function for future import of the same or similar data sets.

Within MATLAB using structures and table arrays can simplify data manipulation and help you to keep track of variables.

Missing numbers are represented in MATLAB as **NaN**. You can choose to omit missing numbers from statistical calculations or replace the numbers with estimated values prior to further analysis.

MATLAB® as a Programming Language

C. R. Webb, M. Domijan, *Introduction to MATLAB® for Biologists*,
Learning Materials in Biosciences, https://doi.org/10.1007/978-3-030-21337-4_11

What You Will Learn in This Chapter

While MATLAB® can be just used as an elaborate calculator and a plotting program, to get the most of MATLAB, you will need to be able to do some basic programming. In this chapter we introduce five key programming concepts which will enable you to write scripts that automate repetitive tasks: relational operators, logical operators, conditional statements, for loops and while loops. This chapter is aimed at readers who have little or no previous programming experience. If you have no experience of coding, these concepts may initially appear a little daunting. However, just like learning a new language, the more you practice and try to incorporate the methods into your own work, the quicker you will reach a point where it becomes part of your natural language.

11.1 Programming in MATLAB®

MATLAB is a structured programming language which means that you can write sophisticated code with relative ease. You do not need to compile your code or declare variables, and there is a vast library of ready-to-access functions which you can incorporate in your code. This chapter introduces the basic coding techniques required to make optimal use of the MATLAB programming environment: logical statements, conditional statements and loops. To understand the content of this chapter, you should copy each section of code into a script file and explore what happens if you make changes to the code.

11.2 Logical Statements

11.2.1 Relational Operators

If you have worked through almost any of the other chapters in this book, then you will have already encountered some relational operators. They follow standard mathematical notation to define the relative values of any two numbers. The six operators are given in Table 11.1.

Table 11.1 Relational operators: fill in the logical value you expect each command to produce (0 or 1) for the values of a and b given in columns 3, 4 and 5. Check your answers in MATLAB®

Operator	Definition	a = 2 b = 2	a = 4 b = 2	a = 1 b = 5
a == b	a is exactly equal to b			
a > b	a is greater than b			
a < b	a is less than b			
a <= b	a is less than or equal to b			
a >= b	a is greater than or equal to b			
a ~= b	a is not equal to b			

Note the operator for 'exactly equals to' requires **two** equal signs. This is because if you use one equal sign, then MATLAB will attempt to assign the first value to the second value (which makes no sense if the left hand or both sides are numerical values) and an error message will be generated.

Relational operators are examples of logical statements. Logical statements, which are true, are assigned the value 1, and logical statements which are false are assigned the value 0. The result of a logical statement can be assigned to a user-defined variable: this new variable will be automatically assigned class type **logical**. To display the class type of the variables in the current workspace, click on the pull-down menu in the Workspace Window, and select **Choose columns > Class**.

Try the following example:

```
A = 2; B = 3;
A==B
Test1 = A==B
Test2 = A<B
Test3 = A~=B
```

Logical values can be added together, allowing us to find the total number of tests that are true:

```
TotalTrue = Test1 + Test2 + Test3
```

We can also find the relationship between every value in an array and some given value. This is useful if you want to rapidly find out how many values in your data set meet a given criteria.

```
MyData = randi(10,[3,4])
TestMyData = MyData<=2
TotalMatches = sum(TestMyData(:))
```

You may be wondering how this compares to the find function we have used elsewhere in the book. The **find** function makes use of relational operators to give the location in the array of any values which are true (the default output is a list of locations by index; see ► Chap. 3 for a reminder of the difference between subscript and index notation). As an alternative to the above code, we could have used the built-in functions **find** and **numel** to determine the number of values in **MyData** that are less than or equal to 2. The function **numel** counts the number of elements in an array:

```
WhichAreTrue = find(MyData <= 2)
HowManyTrue = numel(WhichAreTrue)
```

11.2.2 Logical Operators

Just like relational operators, logical operators result in Boolean output with a value 1 if the logical statement is true and 0 if it false.

Logical operators are often used to combine relational operators. Suppose, for example, that you have a data set in which you want to exclude all values above some lower cutoff and below some upper cutoff. It is not possible to this with a single relational operator. Instead we need to combine two relational operators using a logical operator.

There are four standard logical operators AND, OR, XOR and NOT. The MATLAB notation for each operator and the definition of the operator are summarized in ◘ Table 11.2.

11.2.2.1 Using Logical Operators to Compare Two Values

Type the following code into a script file, and run the code (remember you can run a section of code by pressing CTRL-ENTER or by highlighting the code you want to evaluate and pressing F9). Check that you understand the value output for variables A, B, C, D and E. The quickest way to see the output is to look in the Workspace browser.

```
clearvars
a = 0.5;
b = 2;
A = a==b
B = a < b
C = a >= b
D = a~= b
E = 4.*a == b
```


Now use these logical values in the following logical operations making sure you understand the output:

```
AandB = A & B
AandC = A & C
BandD = B & D
BDandE = B & D & E
ABandD = A & B & D
```

◘ **Table 11.2** Standard logical operators and basic notation in MATLAB®

Operator	Name	Definition
A & B	AND	Both condition A and condition B are true
A | B	OR	Either condition A is true or condition B is true, or both conditions A and B are true
~ A	NOT	Condition A is not true (notice this operator can only be applied to a single logical statement)
xor(A,B)	XOR	Either condition A is true or condition B is true, but conditions A and B are not both true

Repeat the above commands but replace & (AND) with | (OR).

The **xor** (exclusively or) command does not have a shorthand, and we need to use the built-in function **xor**. This function can only be used to compare two values at a time:

```
xor(A,B)
xor(B,D)
```

The ~ (NOT) function only works on single values; however, it can be combined with other logical operators:

```
~ A
~A & B
A | ~B
```

11.2.2.2 Using Logical Operators to Compare Values in Two or More Arrays

Logical operators can be used to efficiently compare two or more arrays. The arrays must be the same size. Suppose, for example, you have collected data on the number of beetles found in each of six locations at two different time points, **Time1** and **Time2**:

```
clearvars
Time1 = [6 12 4
    3 9 7];
Time2 = [4 8 2
     1 3 11];
```

Suppose now we want to identify the locations in which at least 5 beetles were found at each time point. One way to do this is to start by transforming each matrix into a set of logical values by using a relational operator.

```
Time1GE5 = Time1 >= 5;
Time2GE5 = Time2 >= 5;
```

Before you continue, check the output of **Time1GE5** and **Time2GE5**. We have created two matrices which contain logical values. You can check this by typing:

```
islogical(Time1GE5)
islogical(Time2GE5)
```

Compare the output of these two statements with the output of:

```
islogical(Time1)
islogical(Time2)
```

To identify the locations in which at least 5 beetles were found at both time points, we can now use the logical operator & (AND):

```
AtLeast5Beetles = Time1GE5 & Time2GE5
```

We could have achieved the same result in fewer lines:

```
Always5orMoreBeetles = Time1 >= 5 & Time2 >=5
```

Now use the other logical operators on the matrices **Time1GE5** and **Time2GE5** to find:
- All the locations in which at 5 or more beetles were found at least once
- All the locations in which at least 5 beetles were found only on one occasion

11.3 Conditional Statements: if, elseif, else

In our day-to-day lives, our actions are consciously, and subconsciously, based on a series of logical statements. If some condition is met, then we take one action; otherwise we may take no action or take an alternative action. For example, when your pet demands food then *if* they are due some food, you will give them some, but *if* they have only just eaten, you will tell them to wait a while. Computationally we can write these decisions in the form of **if** statements.

```
x = rand();
if x < 0.5
    disp([num2str(x,2),' is less than a half'])
end
disp(['The random value was ',num2str(x,2)])
```

The keyword **if** must be matched with an **end** statement. If you rerun the above code a few times, you should notice that it only displays an output stating the number generated and statement that it *is less than a half* when the condition is true; otherwise it will move on to execute the next bit of code. You could also write this as a function in a separate file (see ▶ Chap. 9):

```
function TestOPt5
x = rand();
if x < 0.5
    disp([num2str(x,2),' is less than a half'])
end
disp(['The random value was ',num2str(x,2)])
end
```

Suppose now we want something else to happen if the logical statement is false – to do this we add an **else** statement. For example, write a new script or function:

```
% function DiffOpt5 % uncomment if you want to make a function
y = rand();
if y < 0.5
    diff = 0.5-y;
    disp(['The value of y is ',num2str(y,2)])
    disp(['y is ',num2str(diff,2),' less than 0.5'])
else
    disp(['y is ',num2str(y)])
end
% end % uncomment for a function file
```

More options can be added in using the keyword **elseif**. Try the following example (again you can write it in a function):

```
z = 100.*rand();
if z < 30
    disp(['Test score = ',num2str(z,3),'% fail'])
elseif z <70 & z >= 30
    disp(['Test score = ',num2str(z,3),'% pass'])
else
    disp(['Test score = ',num2str(z,3),'% distinction'])
end
```

You can also add more than one **elseif** statement if necessary. Copy the following into a new script or function, and run the code several times:

```
Height = randi([150,200]);
Weight = randi([40,100]);
BMI = @(Height,Weight) Weight./((Height./100).^2);
PatientBMI = BMI(Height,Weight);
disp(['Height: ',num2str(Height),'cm Weight: ',num2str(Weight),'kg'])
Advice = @(Statement) disp([Statement,': BMI = ',num2str(PatientBMI)]);
if PatientBMI < 18.5
    Advice('Underweight')
elseif  PatientBMI >= 18.5 & PatientBMI < 25
    Advice('Ideal weight')
elseif PatientBMI >= 25 & PatientBMI < 30
    Advice('Overweight')
elseif PatientBMI >= 30 & PatientBMI < 40
    Advice('Obese')
else
    Advice('Morbidly obese')
end
```

Use the following code to explore how an `if` statement is evaluated when more than one condition is met:

```
mice = 15;
if mice > 0
    disp('at least one mouse caught')
elseif mice > 12
    disp('at least 12 mice caught')
end
```


Once a condition is met, the remaining conditions are not tested, and alternative actions will not be performed even if more than one condition is met.

11.3.1 Nested if Statements

Nested `if` statements allow you to test one condition and then test another condition if the first condition is true. You can nest any number of if statements, but each statement requires its own `end` keyword. In this example the code outputs a course of action depending on whether the patient has a rash and their temperature:

```
RashOption = {'Yes' 'No'};
Rash = RashOption(binornd(1,0.2)+1);
Temperature = randi([36,42]);
FormatText = 'Temperature = %d deg C. Rash present = %s';
sprintf(FormatText,Temperature,Rash{:})
if strcmp(Rash,'Yes')
    if Temperature > 40
        disp('Treat immediately')
```

```
        else
            disp('Check again in 4 hours')
        end
    else
        if Temperature > 40
            disp('Check again in 12 hours')
        else
            disp('Send home')
        end
    end
```

The above section of code could also have been written without nesting by using multiple else statements and the **&** operator to check for each of the four possible scenarios (High Temp & Rash, High Temp & No Rash, Low Temp & Rash, Low Temp & No Rash).

There are often several ways in which you can write code to achieve the same outcome. While some approaches may be more computationally efficient, it is not always predictable which will be faster, in part due to the way MATLAB has been optimized. It is best to follow the advice in the MATLAB documentation (filed under 'Performance and Memory'):

> Write your code to be simple and readable, especially for the first implementation. Code that is prematurely optimized can be unnecessarily complex without providing a significant gain in performance.

You might also want to try rewriting the above code as a function which takes as inputs **Temperature** and **Rash** rather than generating a random temperature and rash status (see ► Chap. 9 for more information on how to write functions).

Suggested Areas for Further Exploration

Two other useful built-in functions which can be used to extend the power of logical operators are **all** and **any**. These functions enable the user to apply the logical operators AND (**all**) and OR (**any**) to each row or column of a matrix. For further details and examples, see the MATLAB documentation.

You may also want to explore the logic operator, **isa**. This operator allows you to question the data type of a variable of interest. You can find a list of all the functions which are used to detect whether something is true or false by searching the documentation for **is***., i.e. type into the Command Window:

```
doc is*
```

If you need to increase code efficiency, then for some logical operators, you can use short-circuit operators, e.g. && instead of the operator & and || instead of |. These can be more efficient than using standard operators as they do not proceed to testing the second condition if the first one is not met.

Another type of conditional statement which can be used to control program flow is a `switch` statement. You can find out more about this type of statement in the MATLAB documentation.

11.4 Loops

Loops enable repetitive tasks to be performed without having to write out the commands each time. We will explore the two main categories of loops:

- A `for` loop is used to repeat a task a predefined number of times.
- A `while` loop is used to repeat a task until some condition is met.

To understand the difference between `for` loops and `while` loops, consider the following example. Suppose there is a field of wheat which you want to check for signs of stem rust. You could use one of two strategies:

- ***Strategy 1*:** `for` strategy, choose a random plant and check for symptoms; repeat 100 times.
- ***Strategy 2*:** `while` strategy: choose a random plant and check for symptoms; stop checking plants when you find one plant with symptoms.

Assuming we do not care about the prevalence of disease, just the presence or absence, then the `for` strategy has the advantage that you know that you are going to check exactly 100 plants. The two disadvantages are (i) if you find an infected plant early on, you still have to check the remaining plants; and (ii) you may misdiagnose the field as disease-free as the disease symptoms are present but not in any of your sample.

The `while` strategy guarantees that you will eventually find the infected plant ***if there is one in the field***, and if you find one after only checking a few plants, you will have done less work than the `for` strategy. However, if the field is disease-free, you will have to keep checking until someone comes to stop you (equivalent of pressing Ctrl+C or Ctrl+Break to stop execution of your MATLAB code). This highlights the danger of a `while` loop – getting stuck in an infinite loop. To avoid this in the field situation, we might keep a record of which plants we've already checked and added a clause to break out of the loop once we've confirmed that every plant is disease-free, i.e. while we haven't found disease and while there are still plants to check. The next two sections introduce the use of `for` and `while` loops in MATLAB.

11.4.1 for Loops

A `for` loop enables repetition of a set of commands for a predetermined number of iterations. You can include as many commands as you like within a loop. The end of a loop requires an `end` statement. The first line of a `for` loop includes a sequence of values through which you want the loop to iterate. If you just want to repeat a loop a specified number of times, n say, then the most straightforward approach is to create a sequence of integers from 1 to n.

```
for i = 1:5
    disp('I love MATLAB')
end
```

The above section of code should have resulted in the text `'I love MATLAB'` being printed to the command window five times. Any sequence of five values will give the same result.

```
for NS = [8 56 8 -2 4]
    disp('I love MATLAB')
end
```

We can also loop through the values in a cell array:

```
for NS = {'a lot','it is great'}
    disp('I love MATLAB')
end
```

The number of elements in the sequence gives the number of times the commands in the loop are executed. The loop runs through each value of the sequence moving from left to right. The current value of the sequence can also be used in the loop.

```
for i = 1:5
    i
end
```

Another example using cells:

```
for NS = {'a lot','it is great'}
    fprintf('I love MATLAB %s \n',char(NS))
end
```

The value in the sequence does not need to be the same as the loop number.

```
L = 0;
for j = 14:-3:0
    L = L + 1;
    disp(['The loop number is ',num2str(L)])
    disp(['The value of j is ',num2str(j)])
end
```

When MATLAB executes the above **for** statement, it iterates through the vector, [14 11 8 5 2]. The parameter `j` will take on each value of the vector in the order they appear in the vector, so on the first run through of the loop `j` has the value `14`, the value of `j` stays at `14` until the end statement is reached. The code then goes back to the top of the loop, and `j` takes the next value, so `j` is now `11`, and the statements in the loop are executed with `j = 11`. This continues until there are no more values left in the sequence. At this point the execution leaves the loop and moves on to the next line of code.

Within a loop the current value of the sequence is assigned to the name you have given the sequence, and it can be used like any other parameter in the workspace. Try out the following example:

```
for x = 1:3
    x.^2
end
```

For increasing integer sequences, the current value of the sequence can also be used as an index for some other variable.

```
for g = 1:5
    gcube = g.^3
    n(g) = g
end
```


Note that in the above examples we have not suppressed the output using the semicolon command since it is useful to see the output of each iteration of the loop. However, it is generally advisable to suppress output of individual loops to speed up run time and to avoid clutter in the Command Window. The next bit of code combines creating a new array with using an arbitrary sequence of values.

```
sqrdata = zeros(5,2);
n = 0;
for mydata = [56 4 3 2 4]
    n = n+1;
    sqrdata(n,1) = mydata
    sqrdata(n,2) = mydata.^2
end
```

The first line of the above code creates a new variable called **sqrdata** which is a 10 by 1 array of zeros. This line can be omitted, and the loop will still work. Unlike many programming languages, in MATLAB you do not need to pre-allocate variables (i.e. declare in advance the size of the **sqrdata** variable). Failing to pre-allocate variables can however slow down the run time for your code.

Below are some more examples of how **for** loops can be used to create new variables. Before you run each section of the code, it is useful to think about what output you expect it to produce.

```
%% Example 1
clear all
n(1)=2
for i = 1:4
    n(i+1) = n(i).*n(i)
end
%% Example 2
clear all
n(1)=0
for i = 1:4
    n(i+1) = n(i)+12
end
%% Example 3
clear all
n(1,1)=0
for i = 1:4
    n(1,i+1) = n(1,i)+12
end
%% Example 4
clear all
n(1,1)=0
for i = 1:4
    n(i+1,1) = n(i,1)+12
end
```

Do you notice the difference between examples 3 and 4? Example 3 produces a row vector, while example 4 produces a column vector.

As with built-in functions and conditional statements, we can nest **for** loops. In the example below, we use nesting to create a times table.

```
TimesTable = zeros(12);
for i = 1:12
    for j = 1:12
        TimesTable(i,j) = i.*j;
    end
end
TimesTable
```

While, it might be tempting to write everything in loops, many tasks can be achieved in far fewer commands by making use of the matrix structure underlying MATLAB. You may recall that in ► Chap. 6, we created a times table grid using only one line of code. Compare the above code with:

```
QuickTimesTable = [1:12].*transpose([1:12])
```

Experimental protocols frequently include loops. Suppose, for example, you are writing instructions to set up a serial dilution to estimate the number of bacteria in a milk sample. The first step is to prepare ten test tubes by adding 9 ml of sterile water to each test tube. You could give a separate instruction for each tube "Add 9 ml of sterile water to the tube at the left end of your rack, now add 9 ml of saline to the next tube, now add ....". We can write the process of filling the tubes as a **for** loop:

```
% create 10 empty test tubes
clearvars
TestTube = zeros(10,1);
for TTid = 1:10
    TestTube(TTid,1) = 9;
end
```

The above for loop was for illustrative purposes – we could have achieved the same result with the verbal command "fill all the test tubes with 9 ml of sterile water" or in MATLAB terms:


```
TestTube1(:,1) = 9.*ones(10,1)
```

Next, we need to label the test tubes from −1 (indicating a dilution of 1 in 10 which can be written as 10^{-1}) down to −10 (dilution of 10^{-10}) in the order that they are on the rack. For each test tube, we will list the dilution index in the second column of the matrix TestTube:

```
for TTid = 1:10
    TestTube(TTid,2) = -TTid;
end
```

Again, you could have achieved the same result in MATLAB in one line:

```
TestTube1(:,2) = -1:-1:-10
```

Now we want to explain why the dilution in each test tube can be written as powers of 10. The first test tube has 1 ml of milk and 9 ml of water. The total amount of liquid in the test tube is 10 ml so the dilution of milk is 1/10. Now we add 1 ml of this diluted solution to the next test tube so the amount of milk in the next test tube is one tenth of 1 ml divided by 10 ml. i.e. 0.1/10 = 1/100 and so on.

```
TestTube(1,3) = 1/10; % dilution in first test tube
for dilute = 2:10
    TestTube(dilute,3)=TestTube(dilute-1,3)./10;
end
```

We can now work out the expected concentration of bacteria in each dilution for any initial concentration and plot the results:

```
InitialConcBacteria = 3.134e12 % 3.134x10^12
TestTube(:,4) = TestTube(:,3).*InitialConcBacteria
% plot
figure1 = figure;
axes1 = axes('Parent',figure1);
loglog(TestTube(:,3),TestTube(:,4)) % do log-log plot
set(axes1,'XDir','reverse') % this forces MATLAB to plot axes from
% largest to smallest value
xlabel('Dilution')
ylabel('Expected number of bacteria in 1 ml')
```

In the next example, we demonstrate how you can concatenate text to save the output of each iteration of a loop to a different file. This example uses a table array so that the data generated can be stored with column names. The function **writetable** is used to output the data to a file.

```
% data set for 4 mice each of which contains 5 rows of time point data
MadeUpData = randi(20,[20 3]);
for Mouse = 1:4
    StartRow = (Mouse-1).*5 + 1;
    EndRow = Mouse.*5 - 1;
    MouseData = MadeUpData(StartRow:EndRow,:);
    % create a table for storing calculations
    MouseResults = table;
    MouseResults.time = [1:size(MadeUpData,2)]';
    MouseResults.mean = mean(MouseData,1)';
    MouseResults.min = min(MouseData,[],1)';
    MouseResults.max = max(MouseData,[],1)';
    MouseFile = ['StatsMouse', num2str(Mouse),'.dat'];
    writetable(MouseResults,MouseFile)
    clear MouseResults
end
```

Let's look at one more example which uses a **for** loop to find any alignments of a short DNA sequences within a longer sequence. The output is the starting position in the longer sequence at which the two sequences align:

```
DNA1 ='ATCGCTTGATATACG'
DNA2= 'TATA'
BaseCount1 = numel(DNA1);
BaseCount2 = numel(DNA2);
Diff = BaseCount1-BaseCount2;
for i=1:Diff
    DNA1part=DNA1(i:i+3);
    Position(i)=strcmp(DNA1part, DNA2);
end
AlignLoc = find(Position)
```

The **strcmp** function compares subsections of the first DNA sample with the whole of the shorter second DNA sample. The function outputs a 1 if the two strings match otherwise it outputs a 0. When the loop is complete, the function **find** is applied to output the starting position(s) where a match was made. A similar function, **strcmpi**, can be used if we want the comparison to be case insensitive.

11.4.2 while Loops

Suppose now that we have a block of code that we want to keep on rerunning until some specific condition has been met. To do this we can use a **while** loop. Copy the following code into a script (or create a function), and run it several times:


```
clearvars
n = 20;
while n > 5
    n = randi(20);
    disp(['n = ',num2str(n)])
end
```

The set of commands between **while** and **end** statements are referred to as the 'body of the loop'. You should notice that all output values of **n**, except the final value, are greater than 5. The last number is always less than or equal to 5. This is because this code keeps on generating new numbers while the previous number obtained is greater than 5.

Now let's try another example. Sunflower seed patterning can follow a Fibonacci sequence. Counting the number of seeds on a sunflower head in a spiral in either clockwise or anti-clockwise direction will result in a number from the Fibonacci sequence. Note that we have already looked at a script (using **for** loops) to create a Fibonacci sequence in ▶ Chap. 4. In a Fibonacci sequence, the next value in the sequence is calculated by adding the preceding two numbers, and it begins as 1, 1, 2, 3, 5, 8, 13, 21 and so on. We can use a while loop to calculate the sequence of Fibonacci numbers up to some maximum value, say 1000:

```
SeedSpiral(1:2)=1;
n=3;
LastSpiral=SeedSpiral(2);
while LastSpiral <= 1000
    LastSpiral=SeedSpiral(n-1)+ SeedSpiral(n-2);
    SeedSpiral(n)=LastSpiral;
    n=n+1;
end
disp(['Last two values in SeedSpiral:', ...
num2str(SeedSpiral([n-2,n-1]))])
```

When you run this code, you will see that, although we set the condition to enter the while loop as **LastSpiral <= 1000**, the final two values in the vector **SeedSpiral** are 987 and 1597. When **LastSpiral = 987**, the loop condition was met, and the body of the loop was still evaluated giving a new value of **LastSpiral = 1597** which was added to the vector **SeedSpiral**. One way to prevent this is to add a conditional statement which checks the value of **LastSpiral** before adding it to the vector **SeedSpiral**. If the value is greater than 1000, then a **break** statement can be used to exit the loop:

```
clearvars
SeedSpiral(1:2)=1;
n=3;
LastSpiral=SeedSpiral(2);
while LastSpiral<1000
    LastSpiral=SeedSpiral(n-1)+ SeedSpiral(n-2)
    if LastSpiral>=1000
        break
    end
    SeedSpiral(n)=LastSpiral;
    n=n+1;
end
disp(['Last two values in SeedSpiral:', ...
num2str(SeedSpiral([n-2,n-1]))])
```

When creating a while loop, it is important to have a body statement where the value in the conditional statement gets updated; otherwise the loop will be infinite. Consider the following case:

```
n=20;
while n>0
    n=n-1;
    disp(n)
end
```

Predict what would happen if we commented out the line that updates n (to do this use the % sign). If you do accidentally write a program with an infinite loop, you can terminate it by pressing CTRL+C or CTRL+Break on your keyboard.

Remember, if you want to set the condition to exactly equal to a value, you need to use == to indicate "exactly equals to" as in the following example:

```
n = 1;
while n == 1
    n = binornd(1,0.8);
    disp(['n = ',num2str(n)])
end
```

Be careful with using exactly equal as a criterion. MATLAB uses floating point numbers in most calculations, and this can result in rounding errors as described in ▶ Chap. 2.

▪▪ Suggested Areas for Further Exploration

A good way to practice writing code is to spend some time on Cody™ (see ▶ Chap. 5 for more details on how to use Cody). You may also want to look up the following MATLAB documentation for advice on how to improve your code:

```
doc Techniques to Improve Performance
doc Improve Code Readability
```


Take-Home Message

Learning the basics of coding in MATLAB® will make it easier for you to write efficient code to perform repetitive tasks and to manipulate data. The quickest way to become comfortable with programming in MATLAB is to apply it in your own work. You can also develop your understanding by using the many resources available in the documentation and on the MathWorks® website. If you have not yet tried Cody™, then give it a go – many of the examples are designed for MATLAB and coding novices, and the opportunity to see how others solve the same problem can also help you to improve your understanding of MATLAB.

Supplementary Information

C. R. Webb, M. Domijan, *Introduction to MATLAB® for Biologists*,
Learning Materials in Biosciences, https://doi.org/10.1007/978-3-030-21337-4

Index

Symbols

A

B

C

D

E

Q

R

S

T

U

W

X

Y

Z

Printed by Printforce, the Netherlands